THE LESS IS MORE APPROACH TO WINE

CHARLES SPRINGFIELD

THE LESS IS MORE APPROACH TO WINE

Published in the United States by way of
The Editorial Stylings,
an editorial division of
The Life Stylings of Charles Springfield,
New York, NY
www.charlesspringfield.com

ISBN (paperback): 978-0-578-42574-0
ASIN (Kindle e-book): B07S9LP54Y

PRINTED IN THE UNITED STATES OF AMERICA

Art Direction by Anne Woelfel © 2018
Cover photography by Curtis O'Dell, 2018

First Edition May 2019
10 9 8 7 6 5 4 3 2 1

the dedication

This book was manifested out of love – the love of family, the love of friends, the love of life, the love of educating, the love of entertaining, the love of learning and, of course, the love of wine.

the table of contents

: the declaration

Wine is, without question, the love of my life.

Don't get me wrong. I know that seems like a really bold statement to make about a beverage.

I'm not a hermit or a recluse. I love people too. I actually happen to have a lot of special people in my life. But please just rock and roll with me on this topic for a few minutes.

I've had a life-long affinity for and with this juice. And with that type of history, the memories are long and plentiful.

As the better part of two decades crawled by, I can recall having deep and heartfelt conversations over bottles and bottles of wine trying to figure out life with my nearest and dearest friends from the different eras in my life.

Thoughts start to swirl around my head of countless nights sitting around soul searching and dreaming about the future, wondering what profession I would pursue and who I would become.

The delicious memory of smelling Sauvignon Blanc on an apartment balcony under the stars in Paris, France, on a warm summer night still makes me smile so wide.

Then the all-too familiar and bittersweet tastes of goodbyes accumulated over the years from packing up to leave one city after another city after another city in the hopes of permanently landing in New York City.

Like a slideshow, those and many other thoughts play a constant flow of rotating images in my head which make me laugh, cringe and cry.

Does any of this ring true for you?

If wine were a person, I truly believe we would be BFFs (Best Friends Forever). I guess, in a sense, we are BFFs.

And it has rarely let me down.

I just now realized – in writing this – that I've never officially taken the time to articulate just how much wine has meant to me throughout my life.

I guess it's better late than never.

It seems that my affection has only grown stronger during my nearly 10 years in the wine industry in New York City.

That's when wine changed my life.

The more I began to unpack the history and peel back the layers of wine education, the more it left an indelible mark on me.

It gave me a new path. A new profession. My third career.

I could not have imaged that this industry would serendipitously fall into my lap as a part-time, wine sales associate all those years ago. Then, nearly a decade later, result in me writing and publishing my first wine education book.

It seems I found my calling. I found my passion. I found my voice. I found my "why."

Did I happen to mention how much I love wine?

In fact, the word *love* might not even be the proper verb to do my feelings justice.

Maybe someone before me has described this feeling better.

Maybe a powerful poet with a profound pen.

Maybe Prince – may he Rest in Purple Power!

There is a lyric from his song "Adore." It seems to provide a better perspective for how I actually feel about wine.

In the song, he exuberantly describes how the word "love" is just *too* weak to properly describe his feelings. *Love* just wasn't enough. He felt the word "adore" was much more fitting.

I adore wine. It's true.

I give it my heart. I give it my mind. I give it my body. I give it my time.

As long as I can breathe, talk and successfully sip on wine – on my own or with the help of a trusted caretaker – I'm going to continue my personal mission of bringing people closer to wine through my wine education work.

My hope is for everyone to understand, appreciate and adore wine just as much as I personally do.

: the introduction

Wine makes occasions social, special and spontaneous.

It emotionally and spiritually connects us to what's occurring in the moment and what's happening in the glass.

Each glance, sniff and sip allow us to commune with nature, the fruit, the region and a history that spans thousands of years.

But let's be completely honest! Wine can also be very overwhelming in terms of shopping, ordering or just casually talking about it.

It is crazy delicious, crazy fun and crazy confusing.

As a wine educator and certified sommelier, I see people get perplexed often and I absolutely understand the problem.

Take a trip to your local wine shop – large or small – and I'm sure you will be greeted with a barrage of unfamiliar brands, grape varietals, styles of wine and varied vintage years.

Since the juice is so damn delicious, you put yourself through the torture of skimming pages of wine lists at restaurants. You aimlessly walk around store shelves hoping something will call out to you. With phone in hand, you use your favorite wine APP to scan bottles for quality assurance *and* the reassurance that you'll happen to stumble across something that might be the right fit for you.

It's enough to give even the most Zen person a panic attack.

I blame it a lot on wine's intimidation factor.

For those of us who happened to grow up in the U.S. – or other countries without a classic wine culture – wine wasn't something most of us grew up drinking. We typically grew up drinking milk, juices and soft drinks at meals or to quench our thirsts throughout the day.

That has put the majority of wine drinkers in this country at a disadvantage compared to our foreign counterparts found in France, Italy, Spain, Germany and other regions around Europe.

In terms of being comfortable with wine, those cultures have no qualms or taboos about children, teenagers and young adults enjoying wine as they grow up.

As a result, there is quite the learning curve when it comes to understanding wine for a lot of us as we approach and surpass legal drinking age.

While we can't turn back time, we can start to turn the situation around in our lives in this very moment.

With the proper information, wine lovers and new wine consumers can learn to talk about, order and shop for their wine with confidence. And they can enjoy the process too.

That's where this book and I come into the picture.

From years of teaching hundreds of wine classes, hosting events and launching wine programs throughout the Greater New York City area, I have crafted an approach in which people can understand wine in real, easy and tangible ways.

I've literally witnessed the "ah ha" moments happen right before my very eyes. Therefore, I've paired that teaching approach with several questions from students over the years and packaged it all into a concise written guide.

I also wanted to capture some of that energy and fun from the classes and channel it into this book! By keeping wine fun, I believe readers are more inclined to continue on their wine education journey – consuming all the various information with excitement and enthusiasm.

I firmly believe the right education can lead to great knowledge. That knowledge can then lead to great respect. And, ultimately, that respect can lead to a greater appreciation of wine.

: the scope of the book

Understanding the complexities of this vast world of wine can be made simple. But it *will* take some work on your part.

Don't worry, though. This is the "less is more" approach, remember!

But how can that be? Less is less, right?

No, not always. At least not in my book.

When developing this book, I was really inspired by minimalism. The concept of putting more value on simplicity and brevity as opposed to complexity and lengthiness really intrigues me.

I especially admire minimalism or the "less is more" approach when applied to other major passions of mine like art, architecture, fashion, music and journalism.

The "less is more" concept was essentially the motto of Ludwig Mies van der Rohe who was a big proponent of minimalism in architecture. He created a design style that represents clarity and simplicity. And since that perspective parallels with my teaching style, I knew it would the ideal approach and name for this book.

When I see the works of two great minimalist artists – renowned painter Piet Mondrian and fashion designer Coco Chanel – I understand that the "more is more" concept can be too much in many cases. As I sat down to write this book, I had all those great artists in mind.

Like a Mondrian painting, I opted for primary, choice topics to feature in each chapter. I was inspired by the way he used primary colors in his paintings so each piece would come together to create the "bigger" picture. In the case of this book, the objective was to feature specific content and place them in a linear manner. Each chapter should effectively support the overall framework of the book, culminating in a fully realized picture of the wine world.

Then, there is Mademoiselle Coco Chanel. She wasn't a minimalist artist per se, nor did she design men's clothing. Yet I really appreciate her minimalist design esthetic and game-changing concepts. More than that, though, I love Chanel's philosophy and advice about presenting yourself to the world.

"When putting on accessories, take off the last thing you put on," said Chanel.

That spoke volumes to me in terms of editing. Therefore, I applied that principle to writing and editing. I wanted to make sure the book doesn't overwhelm the eye with too many details – defeating the intended purpose by having information lost in an overly busy presentation.

In terms of actual writing, I've been a big fan of this acronym – K.I.S.S. – since I studied journalism in college and worked in the media. That stands for "Keep It Simple, Stupid." Stupid, referring to the writer – of course; not the reader.

The goal in writing for the masses is to communicate in a way that can be easily understood by all. As a journalist, I was encouraged to report the facts correctly, concisely and accurately without over complicating the matter. Therefore, in the newspaper newsrooms in which I've worked, editors would remind us to "K.I.S.S." the story.

I definitely wanted to K.I.S.S. this book. We know all-too-well how confusing wine can be. It was my goal to make this complex and broad topic more straight-forward by stripping out several highly specific details found in many of the all-encompassing wine books.

Don't get me wrong. There is a time and place for those larger volume wine books too. I love them. I have several in my library and hope to get a chance to write one of those books during my career. However, given the sheer size of some of those books and our shortening attention spans, I thought this approach would be successful in getting the overall essence of wine fully consumed.

Make no mistake. There is a lot of information provided in this book. It's pretty concentrated. But hopefully the readers can get a strong foundational understanding of wine here and further enhance his or her knowledge moving forward through other educational resources.

: the way to use this book

The easiest way for me to learn is to have information laid out in a linear manner.

Therefore, that's what I tried to do with the organization of this book – provide enough context from the beginning to the end so information flows into each other fairly logically.

Repetition always helps me too. There will be a series of reoccurring themes throughout the book. One reason is that several elements of the wine world tend to overlap and connect with each other. The other reason is for retention sake. Repetitive themes help drill the information into your memory and aid in overall comprehension.

However, I want you to use this book as you personally see fit. I am a Virgo, so you already know where I'm going with this statement. Virgos can be natural control freaks. That characteristic describes me well in certain situations. But I am in no way, shape or form going to dictate how you read this book. We all learn differently and have different interests.

You can read this book from cover-to-cover. You can jump to the chapters that better suit your immediate needs. Or you can skip around and use the book for quick reference material. All of that is perfectly fine.

Each chapter shares a unique story. The chapters can stand on their own. But the intention of the book is to have each chapter progress into the next for a cohesive, full story.

Ultimately, how you approach this book is very individual. It's just like how one approaches wine. It's very personal.

CHAPTER ONE: **the history**

I have always loved a good origin story.

You know the kind of stories when you find out how Dr. Robert "Bruce" Banner became "The Incredible Hulk." How Anna Mae Bullock became Tina Turner. How Stella got her groove back. Or how a piece of chocolate falling into a jar of peanut butter resulted in the ingredients of one of my all-time favorite candies – Reese's Peanut Butter Cups.

Fiction or non-fiction, these stories give you incredible insight into how a person, place or thing came to be and got shaped over time.

That knowledge gives you something tangible to hold. Something remarkable to respect. Something more to admire. Something deeper to love.

You understand the full story.

Well, when it comes to wine, "the full story" is massive. Its cup runneth over.

Wine has an incredibly varied and vibrant story that spans more than 8,000 years. It is jam-packed with details, way too many to fit into this "less is more" approach.

However, the history is very crucial to fundamentally understanding wine. And ironically when it comes to wine education, the topic of how and where wine originated rarely springs up.

Many publications, documentaries and even wine educators – including myself – tend to start with what wine *is* or how wine is *made.*

We tend to skim over the history.

I totally understand why. We don't have a lot of information about the origins of wine. We have a lot of theories and speculation on the topic. Since wine predated written language, there are no documents to review and there is definitely no one physically alive to interview.

While we don't have all the details, there are a lot of fragmented pieces of information out there in which to thread together a fairly comprehensive story. Or, at the very least, hit some of the major highlights.

This chapter is about providing a larger historical perspective about wine over a few thousand years by excavating untold stories of wine and weaving those together with well-documented occurrences and dates.

But be forewarned. This is the longest chapter of the book. Even with the "less is more" perspective, it was necessary to provide enough context, broken into specific eras, crucial to understanding why things are the way they are in our wine world today.

And this is where we start.

: the beginning

The discovery of wine can be called serendipitous.

It was a slew of happy accidents and unintentional events that allowed the ancient grapes of that time to transform into the intoxicating liquid we now know and love as wine.

Long ago, there was a time in history when the Paleolithic period (Old Stone Age) was shifting to the Neolithic period (New Stone Age). The time spans from before 10,000 B.C. to about 4,500 B.C. by historical documentation.

These early humans in the Paleolithic era were said to be "hunters and gatherers" based on their roles in life at that time. Men did the hunting of animals for food and clothing. The women did the gathering of the nuts, berries, vegetables and firewood. It has been determined to have been a very nomadic culture where humans moved from place-to-place in search of food, water, shelter and overall survival.

Migration took many of these particular "hunters and gatherers" to a particular area of the world, which is referred to as Mesopotamia, the "Land in Between," located between the Tigris and Euphrates Rivers.

Mesopotamia was coined the "The Fertile Crescent," the crescent shaped area along the Mediterranean where noted early civilizations – the Sumerians, Babylonians, Canaanites and Assyrians – habituated this part of the world. The location is where modern-day Iran and Iraq are now located along with Georgia, Armenia, North Africa, Syria and Turkey.

That's also the time and place where the "wine vine" was expected to have been discovered, which is called Vitis vinifera.

These areas are a substantial distance away from most of the highly regarding wine regions of today's wine world. However, you have to keep in mind that the borders and boundaries of countries and regions that we know to exist today, didn't develop until much later in history.

Apparently, there were ancient wild versions of the Vitis vinifera species growing in this region.

In plentiful periods there would be extra food that would accumulate from the women's efforts of gathering. Those food items, including grapes, were probably set aside somewhere for storage.

As nature would have it, the extra supply would spoil. The crazy thing about grapes, however, is they contain all the needed elements – from the outside of the skins to the inside of the grapes – to self-ferment or turn into wine.

We'll dive more into the fermentation process later in the book.

Not immediately knowing what exactly happened to the grapes, the members of the camps would taste the spoiled grapes and feel and taste something different than they were used to experiencing. They would get an intoxicating feeling from the development of the alcohol – probably ranging from calming and cheerful to medicinal and magical.

This discovery turned out to be somewhat of a "happy accident."

Like most accidents, the situation was what it was. No one believes these wines could have been good, taste wise or smell wise.

However, it provided them with a feeling they had never experienced before. That must have felt otherworldly. It must have felt so good, that archeologists suggest that they started letting their grapes spoil on purpose as a regular practice to enjoy the intoxicating mixture – flaws and all.

And that is how wine got its very early and very humble beginnings, somewhere around 10,000 B.C.

: the early stages of agriculture

The Paleolithic era, nomadic lifestyle, started to transition to the Neolithic era, where tribes began to make certain areas their permanent homes.

Once they staked claim on various pieces of land, the new residents got more and more familiar with their surroundings. They learned how to build on it. They learned to cultivate it. They learned to farm it. They learned how to make the land work for them.

Overtime, from about 9,000 B.C. to 5,000 B.C., the civilizations became increasingly sophisticated. They had developed the wheel, the use of fire, the creation of tools, the domestication of animals and early forms of writing that were unique to them. There was also a lot of trial and error when working the land, creating the early stages of agriculture.

What started out as a "happy accident" in wine moved to "purposeful production." It would later blossom and scale into more robust winemaking when grapes were domesticated around 3,000 B.C. That is said to be the genesis of the grapes in which we are familiar with today.

It is important to note here that this book and the global wine world is based on this particular grape specifies: Vitis vinifera.

Other civilizations in different parts of the world during the same time had also discovered agriculture, with some having their own grape species in their own respective regions – the Indo-Gangetic Plain, North China Plain, Central Andes and Mesoamerica. But it was the "Fertile Crescent" that has been reported as the official roots of the vinifera wine grape species.

Now back to our regularly scheduled program.

After much of the focus was off of forging and gathering food for daily survival, members of the townships started to focus their attention on other matters. A paradigm shift started to occur. The tribes began to blossom from small communities to townships to villages to cities and to early empires.

With that happening, they started to create codes of ethics, various responsibilities and common ways of doing things. It wasn't long before wine became a staple in daily life.

Settling down in one location proved beneficial for these civilizations. In addition to growing their communities and learning how to manage them, it also allowed them – probably propelled by wine and other intoxicants – to open their minds up to concepts of religion (polytheism/multiple god forces), communication, astrology, science and trade opportunities.

While there were some dynamic upsides to their new settled lives, it also put them at risk. Other populations who were still moving around these areas for their daily survival, weren't always pleasant visitors when embarking on these established residences.

Some communities found themselves fending off invaders who wanted what they had. Those visitors didn't have any problems using force to get what they wanted.

This mentality unfortunately put the wheels in motion for a long future of ongoing conflicts: crime, war and pillaging for the sake of survival, power and dominance. These are, of course, themes that continue to permeate the world today.

: the spread of the wine world

Life seemed to be everchanging. There were growing civilizations. Fractured civilizations. Falling civilizations. Migrating civilizations. Merging civilizations.

While there were some dire challenges to face – particularly associated with emerging societal ills – it also led to some new opportunities.

The development of communities and cities sparked entrepreneurship, burgeoning industries and trade. Some of these communities found themselves exploring and crisscrossing the neighboring regions to make new allies by trading information, goods or services.

With this ancient version of "information sharing," the growing vine agriculture and winemaking world was able to expand to places that didn't have access to information on grape growing and/or didn't have their own grapes to grow.

It was around this time, between 1500 and 300 B.C., a group of Canaanites (from modern-day Syria) – who later became known as the Phoenicians – began to sail and trade across regions in the Mediterranean from Lebanon, to their established colonies in the region and eventually from their new home in Carthage located in North Africa.

Two of the most famous trading partners and major benefactors of the Phoenicians were the Egyptians and the Greeks.

This is when historians suggest the Phoenicians introduced both civilizations to wine. For the Greeks, they provided grapevines, the knowledge of grape growing, winemaking, wine storage and transportation in amphora clay pots.

With Ancient Egypt, the two civilizations are reported to have shared their knowledge of many disciplines and practices back and forth. The Phoenicians shared and traded everything from cedar from Lebanon for construction, the purple-dyed fabric in which they were famous for in the region and, of course, wine.

Wine was important to life in Egypt, especially for the nobility and rulers of the time. Well-known for their Pharaohs and Gods, Egyptians worshipped the God Osiris – the God of wine, afterlife and rebirth. With much of the wine at that time being red in color, it resembled the life force of blood. Therefore, it was very important to Egyptians in both life and in death.

During that time, wine was made, exported and shipped from place-to-place throughout the region. Fermentation vessels consisted of amphora clay pots in the ground and above ground in the different regions, some of which are still in existence today. In select parts of the country of Georgia, you can still find ancient Kvevri vessels that were used for winemaking.

Wine wasn't a major interest for the Greeks initially. But they slowly started to incorporate wine into their daily lives, eventually putting a new emphasis on vine agricultural practices for several reasons. With their growing power and influence – intellectually, artistically and with their military – wine became a national symbol of economic strength to the city-states and the lands they colonized.

Greek wine was soon said to have become all the rage at that time across the regions. The wines of those days still had a long way to go before they would evolve into what we drink in our glasses, however. Much of the wine back then was mixed with herbs, sea water, spices and other additives to enhance the flavors.

To help preserve the wine from spoiling, many wines were either topped off with olive oil that would float to the top of the wine to help limit the interaction with oxygen in the clay pots. Sometimes winemakers would coat the amphora pots with pine resin to help seal the wine vessels. Some wines called Retsina are still produced with the pine resin in Greece.

The wine culture was also spread by the Greek's polytheistic religious practices. It became central to the growing religious and spiritual rituals in Greece. They worshipped and honored the Greek God of wine, Dionysus, through many festivals and celebrations in Athens. They also honored the other Greek Gods who were believed to have resided on Mount Olympus.

The intoxicating effects of wine opened the Greeks up to new planes of thought, which can probably be attributed to their scientific theories, artistic endeavors, astrology, architecture, medicine and theological philosophies. That includes inspiring the works and philosophies of Socrates, Plato, and Aristotle to name a few.

As wine became more of a sought-after product from Greece, it became a suitable offering to gods and kings in nearby regions. The Greeks eventually shared its new winemaking techniques with Egypt and areas that are now Spain and Southern France.

And as fate would have it, this ancient wine world became magnified when the Greeks spread their wine expertise to a region in Southern Italy.

In addition to the wine culture that was led by the Etruscans in the northern part of Italy, the seeds were being planted throughout the region for a major new era in wine.

: the rise of a super power

Everything must change!

When the next superpower civilization came into existence – the likes of which the world had never seen before – it changed the wine game forever.

Enter the Ancient Roman Empire.

There was no larger, more powerful or more dominant force than the Roman Empire at the time of ancient civilization.

Let's check the receipts.

The empire grew to rule over approximately 60 million people over 500 years. It covered about two million square miles. That area covered what is now Europe, The Middle East, North Africa and parts of Asia.

Augustus Caesar was the first emperor in the year 27 B.C. Rome was the initial capital, ultimately becoming the largest city in the world at that time. It was expected to have housed about one million residents.

This empire has left a lasting legacy on today's western cultures. The list of contributions is substantial, including everything from social services, city infrastructures, legal systems and a strong military.

As the Roman Empire continuously expanded by military force, it absorbed the existing cultures, practices, traditions and beliefs of the people and places it dominated and conquered.

Wine was no exception.

Before the empire started, the city of Rome was under Etruscan rule and Greek wine was the gold standard. Greek wines were traded at higher levels, while wines made under Etruscan oversight were less prestigious and lower in demand.

Wine became a major industry under the empire. It was taken extremely seriously for supplying their growing populations and for trading with other regions.

The winemakers under the Roman empire would take the learnings from Greek, Etruscan, Phoenicians and Egyptians winemakers – vine farming, plot selection, winemaking, transportation and storing concepts – and morph those into new wine practices and priorities.

These early winemakers worked to fix issues they were facing with wine by creating new technologies, consistently rewriting the playbook and elevating the entire concept of wine to another level.

It was also during this time when the winemakers started to look at land in a more serious way. They began pondering and experimenting with the effect that the land, regions and climate had on certain grapes.

They also started to move away from the use of amphora pots and moved into barrel and glass bottle production. They started expanding on the concept of the wine presses. In addition to using foot treading, they found that presses were able to extract juice from a larger number of grapes to increase the production of wine and make different styles.

Other advances in wine were very important during this time to supply its high demand. In the Roman Empire, wine consumption was more democratized – at least more than in Greek and Egyptian societies. Wine was reported to have been consumed by everyone from nobility to servants. It was used to get soldiers ready for battle. Actually, historians suggest that Roman soldiers consumed about a pint of wine a day, which is about 16 ounces, and up to two pints, or 32 ounces, before heading off to war.

Wine was a part of sporting events. Gladiators consumed their fair share of wine before battles. And it was suggested to be sold in the local taverns and in specialty wine shops.

Then there were the Gods – especially one in particular: Bacchus, the Roman God of Wine and Revelry.

Much like a lot of the earlier civilizations, most of the Roman Empire practiced polytheism.

However, there was Judaism being practiced by Jewish citizens living in various parts of the empire. Jews are an ethno-religious group and one of the oldest monotheistic societies believing solely in one God. Wine had been a long part of the Jewish culture as well, for about 5,000 years, and used for many sacramental purposes.

But with the polytheistic members of the population, they adopted their own God of wine modeled after the Greek God Dionysus. Like in Greece, there were many celebrations and religious activities throughout the Roman Empire in honor of Bacchus.

As the empire continued to grow, the Romans spread their enhanced knowledge of grape growing, vineyard management and winemaking to other regions under its rule. The wine culture spread to places across what is now Europe, places like Gaul (France), Germania (Germany), Britannia (Britain), Hispania (Spain and Portugal), the rest of Italy and other nearby locations.

While polytheism was the way of the world for the majority of citizens, there were a growing number of people who followed the teachings of a man named Jesus Christ during the 1st century.

Different religious beliefs and theologies tended to be allowed in the empire, like Judaism. But the teachings, and the growing following of Jesus, appeared to be seen as a threat to the government and was punishable by death.

Jesus Christ was ultimately put to death by the Roman Empire around the years 30-33 A.D. in the 1st Century. Despite, or in spite of, the death of Jesus, his teachings were kept alive by his followers, and were spread throughout the empire. His followers continued to grow into a larger monotheistic religion that became known as Christianity and the followers became known as Christians.

Christianity, just like Jesus, was also looked upon as a threat to the Roman Empire. Any followers who were caught practicing the religion were punished and, in some cases, put to death as well.

Then, a few hundred years later, in the 4th century things dramatically changed. Christianity had grown so much throughout the communities and regions in the Roman Empire that it created such a major shift in their belief system. The influence was so strong that the ruling emperor himself, Constantine, converted to Christianity in 312 A.D.

Constantine not only converted, he legalized Christianity in the Roman Empire in 313 A.D. and he decided to create a new Christian capital in the empire. He moved the Roman Empire's capital from Rome in the west to the new Christian capital of Constantinople in the east of the empire.

Catholicism soon became the Roman Empire's official religion. It was seen as the formal continuation of the growing Christian community established by the disciples of Jesus.

The shift in religious beliefs had a major effect on wine consumption under the empire. As polytheism – which was known for decadent over consumption – slowly started to become a thing of the past, monotheistic life had a much different view on wine drinking.

It didn't put an end to the importance of wine, however.

Wine became more engrained in the religious practices of the Catholic church. It became an instrumental part of the religion as it was required for Mass where wine symbolically represented the blood of Jesus.

With such a high regard for wine, representing the blood of Jesus, Catholic monks started to take up and take over the practice of growing grapes, making wine and planting vineyards to create a healthy supply of sacramental wine. As the church grew across the empire, so did wine production.

And then about one hundred years later, the Roman Empire came to its end after 500 years of being in power. The collapse of the empire left about 60 million residents, thousands of miles filled with villages, towns and cities and – not to mention – the wine world hanging in the balance with no clear path in which to move forward.

: the dark decline of the western world

Too big to fail!

That phrase sounds eerily familiar.

Yet fail, indeed, it did.

That was the case for the collapse of the massive Roman Empire – which almost meant the same for the growing wine industry.

Through a series of events, both internal and external, the Roman Empire experienced a long, gradual fall over the course of two hundred years, officially collapsing during the 5th century.

Once the conqueror, now the conquered, the empire was dealt its final blow by a Germanic king in 476 A.D.

The fall of the massive and powerful empire left millions of citizens ungoverned, unprotected and without the established structure they've grown accustomed in which to live their day-to-day lives.

This would be the start of the Middle Ages and, more specifically, the start of what's referred to as the Dark Ages. This period lasted roughly from the year 500 to the year 1,000.

There was still a piece of the Roman Empire in the east that was centered around Constantinople. Later known as the Byzantine Empire, it continued to hang on until the end of the Middle Ages in Medieval times. However, the end of the Roman Empire era left the population fragmented and opened the door to a very chaotic time and a slew of issues.

The collapse of the government meant that there was no one centralized power to govern over the areas to continue social services, keep order among the residents or keep up with the maintenance of the larger cities. That led to some very unsanitary living conditions.

It also left the areas with no military force to protect the regions from the continuous onslaught of nearby Germanic and Romano-Germanic Kingdoms – also known as Barbarians – wanting to profit from the empire's demise.

The unsanitary conditions coupled with the pillaging of the inhabited regions created a mass exodus of residents from the cities and villages to find new places of safety and shelter. This time was also known as the Migration period. This would be the very beginning stages of the regions that would make up what we now know as modern-day Europe.

One aspect of life that became a constant – and saving grace (no pun intended) – in the lives of the people in these regions was the Catholic church.

The Dark Ages marked a major turning point in the role and growing power of the Christian church. During this time in history, the Catholic church had enough land, wealth, power and influence to play a significant role in the survival of people in that part of the world.

The church helped to unite these regions despite the destruction going on at the time. In addition to spiritual guidance, people depended on it for many of the services that were now becoming a thing of the past. It provided some structure during those uncertain times as the residents sought to move forward with their lives.

The monks from the Catholic church, who were now well versed in winemaking practices, helped to keep the wine world alive – for the sake of the spiritual and physical well-being of their congregants and residents.

With no centralized government to manage the cities, counties and rural areas, the quality of life essentials under the Roman Empire quickly started to fade away. Therefore, it wasn't long before the water became polluted and was consisted unfit to drink. Wine became necessary for human survival.

But long gone were the days of lavish parties, excess and revelry. Wine was now more a means of survival. It became a matter of life or death and also a path to eternal salvation.

Yet while most of the wine was made for sacramental purposes in church, there was a fair share of "secular," non-church wine for personal consumption around the cities and townships and for trade purposes.

As society was finding its new footing and trying to rebound around the 7th and 8th centuries, the Germanic rulers started to set up a distinctive, new structure for daily life.

With this new structure, life was going to be very different than in centuries passed. During the Roman Empire, the Germanic population previously inhabited and setup kingdoms in the Northern regions above the Roman Empire's territories. The Germanic population just didn't operate the same way, have the same priorities or necessarily quite share the same beliefs as the Romans.

It can be said, however, that they did have a healthy amount of respect for the Roman Empire's achievements – as well as the societal contributions of the church.

The new Germanic leaders started to convert to Christianity around this time, potentially as a way to up their own game and elevate their standing in society. That's just a small glimpse into how powerful and influential the Catholic Church was in the Medieval world. The new rulers considered joining the church would be a positive move and not a threat to their agenda.

The Germanic rulers eventually set up a feudal system – a social hierarchy system from kings, lords, barons, clergy and peasant classes – to establish a form of protection, security and normality throughout the land.

In this system, wine continued to be viewed as a tool for survival. It was now the preferred beverage to drink in the areas that were becoming Britain, France, Italy, Germany and Spain, serving as a way to honor Jesus Christ. Then it also functioned as a healthy source for hydration, medication and for general socializing.

Around this time, the remaining dust from the fallen Roman Empire finally started to settle at bit. Life, however, was still very much touch and go as these residents started to transition into their new way of life.

The light at the end of the tunnel, unfortunately, was not quite there just yet. In fact, it would take about 700 years before society started to officially experience a thriving rebirth of daily life and a rebirth of the wine culture.

Conversely, that light had already been shining brightly in a nearby area over those same 700 years when areas of Europe were rebounding. That enlightenment would eventually make it to Europe through a series of influences that would have direct and indirect impacts on the wine world.

: the light of hope from Africa

The Roman empire stretched over about two million square miles during its 500 years in power.

Its collapse opened the gates for other bodies of power and invading forces to stake claim on the unprotected, ungoverned areas. That meant the Middle East, North Africa and parts of Asia. That is, of course, in addition to regions in Europe.

With new leadership, different thinking and new ways of life tend to be put into place: new regulations, new interests, new priorities and new principles. That happen with the Germanic forces that invaded several areas throughout the European region.

A slightly different shift was happening in the Arabian Peninsula that would spread from Arabia across North Africa to West Africa and then up north to southern parts of Italy and France, and then up to the Iberian Peninsula of what's now Portugal and Spain.

This period would alter the scope of the world's religious landscape, inspire a renewed interest in "classical" thinking, readjust the way mankind saw its role in the world and help plot out how and where the wine world would spread throughout the planet.

Enter the Islamic Empire.

Around the 6th century in Arabia – a peninsula of Western Asia just northeast of Africa – there was a lot of disorder: politically, religiously and commercially.

That was to be expected after the fall of the Roman Empire.

A wide range of spiritual philosophies existed in this region under the Romans. Polytheism was still very prevalent in the region, Judaism was observed by a portion of the population and Christianity had already been spreading in the northeast, northwest and south of Arabia.

After the empire collapsed, the societies in this region were collectively looking for something more in terms of their spiritual well-being. Maybe it was to make sense of what was going on around them. Maybe it was to shape the fate of their futures. Maybe it was to just have a unifying force again.

Then in the early part of the 7th century, around 610 A.D., a young merchant from the Arabian city of Mecca named Muhammad started telling a small group of people that he had been visited by the Archangel Gabriel. Muhammad suggested the angel revealed the true words of God to him. He ultimately wrote them down to create the spiritual guide and holy book called the Quran. Those texts and philosophies ultimately led to the official foundation of a new monotheistic religion called Islam.

This took place about 600 years after Christianity came into existence. After Muhammad died around the age of 62 or 63, Islam grew despite, or in spite of, his passing just like Christianity did after the death of Jesus. That created a group of Islamic followers called Muslims who worship Allah.

Both monotheistic religions – Christianity and Islam – amassed large followings throughout the region, ushered in new ways to worship under one God respectively and allowed their followers to have some type of access to power and influence.

Christian monasteries were constructed throughout some regions. Muslim mosques were constructed in others. Some residents converted to Christianity and some to Islam. And Judaism continued to be observed as it is an ethno-religious group that people are born into or in which they can convert.

The residents eventually found their respective religions but didn't quite have a unifying governing power. That was until the Islamic conquest took place from east to west, dominating North Africa and select areas to the north in Europe, merging many cultures including the Egyptians and Berners to form a new Islamic empire and Muslim nation.

That dominance was partly made possible by the wealth the Islamic Empire acquired through the mining of gold in Africa starting back in the 7th century, funding wars and spreading the Empire and its beliefs to new places.

Wine, of course, had been a big part of the Christian and Jewish faiths. It was also a major part of life in many of these conquered regions prior to the Roman Empire. In Lebanon, for example, there had been a long and stable wine culture started by the Phoenicians.

Throughout the Arabian Peninsula, wine was also traded by Arab merchants to many of the areas that were not suitable for vineyard development and winemaking from grapes.

In contrast, many of the areas were now under Muslim law after the conquests in the 7th and 8th centuries. That consequently altered the religious, political and cultural landscape of North Africa and the surrounding regions in the empire.

Under Muslim law – as stated in the Quran – dietary restrictions forbid the consumption of wine and other alcoholic beverages. This created a major shift on the importance of wine for many of the empire's inhabitants.

But it didn't change it for all.

People that converted to Islam in the growing empire were to practice Muslim religious law. However, inhabitants who were Jewish and Christian were able to maintain their sacramental wine practices and regular consumption. Therefore, wine was still being traded and produced in some regions like Iraq, Persia and central Asia during this time.

As to be expected, with Muslim religious law in place, wine's existence under this empire would be fairly turbulent.

On one hand, wine seemed to surprisingly thrive in select places throughout the Islamic empire. It was a part of the empire's economy, war customs, written about in poetry and part of the regular lives of many of its citizens. Yet in other places and under certain leaders, the wine culture was eradicated for hundreds of years. A large number of grape vines had been ripped out of several regions under Islamic rule. And if the vines were permitted to stay, they were only allowed to produce table grapes or raisins.

: the Golden Age of Islam

While the disjointed regions of Europe had been experiencing some dark times, many areas in the Islamic Empire were experiencing a Golden Age.

It was all sparked by the spiritual texts of the Quran.

The Islamic holy book promotes the pursuit of knowledge. The teachings encourage a strong commitment to learning, seeking information and encouraging literacy among all its followers.

During the period between the 7th and 13th centuries, becoming educated was in stark contrast to what was going on in Medieval Europe. In Europe, illiteracy was a pretty common and accepted way of life for most of the population. Only the higher classes and clergy were taught reading and writing skills.

With education being very important to the Islamic empire, a series of education centers, also known as madrasas, were constructed.

There were two prominent education centers under the empire. The first was the University of Al Quaraouiyine. It was founded in the 9th century in Fez, Morocco and is referred to by scholars as the oldest university in the world. And later there was an educational institution in the ancient city of Timbuktu located in Mali in West Africa. It became well-known for the teaching associated with three mosques in the city from the 12th to the 17th centuries.

By the 9th century, a major emphasis was on science, math, astrology, medicine, art, agriculture, history and literature. That led to a growing population and pipeline of scholars, scientists, doctors, artists and educators.

That quest for knowledge created a strong connection amongst the diverse group of citizens in the empire – forming an information exchange and collaboration between Muslims, Christians and Jews.

There was a big focus on Greek scholarship at these educational centers. The residents were interested in the classic teachings stemming from Ancient Greece. Therefore, these scholars from the different faiths worked together to take the ancient works of Parmenides, Pythagoras, Socrates and Plato and translate them into Arabic to be housed in the education centers for people to study.

And it was not just for people within the city limits to use for research and knowledge. This wealth of knowledge attracted scholars and intellectuals from all over the regions – from the East to the West and North – to study in their numerous libraries.

Timbuktu had developed an unparalleled reputation as a major educational center of the Muslim world in the 12th century, which was funded by its thriving economy. Their trading and selling of salt, gold, spices, textiles, dyes and other goods allowed the educational center to conduct research and acquire ancient scholarly works from the Byzantine Empire. Those trading practices also helped pass their passion for knowledge along to new trade partners and areas they conquered.

As the Islamic empire stretched across North Africa and up into the Iberian Peninsula around the 8th century – Portugal and Spain – it created a lucrative Trans-Saharan trade route and a strong cultural bond between the two continents: African and Europe.

The regions of Spain and Portugal were under Islamic rule for about 700 years, ruled by the Moors, Muslims from North Africa who were later referred to as the Spanish Moors.

This put the African treasures of gold from West Africa and salt from the Sub-Sahara trade routes at the forefront of desired products, helping the Islamic empire to spread wealth, luxury, architecture and educational centers throughout Europe.

The quest for education spread as far and as fast as the European's population's quest for gold. That ultimately led to a new Islamic capital city and educational center in Andalusia, southern Spain, called Cordoba.

Like its educational center predecessors in Africa, Cordoba became a legendary learning center in Spain. It was famous for its multitude of libraries and contributions on architecture, medicine and math. In Cordoba there was the expanded use of the Arabic numerals that later replaced Roman numerals, as well as an amplified interest in ancient Greek and Roman life through literature and historical documents.

This Spanish city helped to continue the Islamic tradition of educational collaborations between the Muslim, Christian and Jewish populations, fostering the same type of mutually beneficial relationship of sharing information like in Fez and Timbuktu.

While this seemed like a very utopian existence, there were still very strong and differing schools of thought in terms of religious beliefs and power dynamics.

Those theological points of differences between Christians, in the nearby northern areas in Spain, and the Islamic Spanish Moors led to some major conflicts. Both sides were driven by the mission to place the world they governed under the worship, belief and principles of one true God – just by using different approaches.

The fallout of those differences – along with the goal of expanding their own empires and religions – led the two sides down on a long path of dark, deadly destruction. Wars were fought in the name of "God" during the Crusades that were initiated by the Catholic church in 1095.

Those conflicts would last for about 400 years. By the year, 1492, the Spanish Moors who had ruled Spain for the last 700 years were pushed out of the country, putting an end to Islamic rule on the Iberian Peninsula.

While the Islamic Empire was retreating out of Europe, the gold, educational scrolls, Arabic numerals and a new way of thinking were already a part of European life. Those elements all started to make their way from Spain across Europe and into Florence in northern Italy, the former home of the Etruscans.

Change was inevitable. Fueled by a renewed focus on intellectual expression and funded by the gold from Africa, Europe was now equipped with the proper tools to break free from its dark period of war, famine and the wide-spread disease of The Black Death, and finally step back into the light.

That light was soon about to shine on the wine world as well, which had been looking at a very uncertain future.

: the renaissance of the wine world

Enter the European Renaissance.

Transformation was all around.

Spain was under new leadership. With King Ferdinand II and Queen Isabella at the helm of the country, Christianity was restored as its national religion by the late 1400s.

While the Islamic Empire was defeated and pushed out of Spain, they had conquered Constantinople in 1453 – the former Christian capital of the Roman Empire. The conquerors changed its name to Istanbul and turned it into the new Muslim capital of the Islamic Ottoman Empire.

Many of the Christians fled Constantinople, now under Muslim rule, and moved west to places like Italy, bringing with them ancient Greek texts, scrolls and Christian Greek Orthodox theology.

The large amounts of gold from Africa helped to increase and expand trade channels, establish wealthy family businesses and create a powerful new banking industry led by the Medici Bank in Florence in 1397. Gold in the form of the florin became the main currency used for "international" trade operations.

With society shifting from its land-based economy under feudalism to more of a money-based economy with gold as the standard currency, urban capitalism became a new way of life. Trade and industry started to be controlled by individuals and private business owners, rather than the "state" controlling the political and economic power of its citizens.

People were migrating back to the west. With more residents and more wealth, the growing populations throughout Europe started calling for a higher demand for wine. That demand started to overflow outside of the purview of the Catholic church and soon into its own independent industry.

The populations were inspired by a new way of thinking about their lives. It was a movement that had been slowly simmering over the last couple hundred years in Europe due to the educational pursuits from the Islamic Empire. Inspired by the philosophies of Protagoras – the ideology that man is the measure of all things – an intellectual movement was ignited throughout regions of Europe that empowered mankind to shape the world in which he or she lived.

With increased wealth spreading around the region and an increased interest in education, the European Renaissance became a dynamic intellectual time with the concept of humanism and expression as its key drivers. This mixture helped further evolve many different forms of expression from art, literature, architecture, theology, innovation and scholarship.

The great thinkers of this time wanted to reclaim the former glory of these fallen civilizations while adding their unique perspectives, big personalities and improving craftsmanship to shape the course of their futures and legacies. This became an important time for the artist, the scholar, the architect, the merchant and the winemaker.

In terms of wine, this period put wine back into an "international" spotlight to the likes of what had not been experienced since the fall of the Roman Empire.

Prior to the fall of the empire, the Benedictine and Cistercian monks had established important wine regions in areas now known as Champagne, Bordeaux, Burgundy and the Rhone Valley in France, and the Rheingau region of Germany.

The church was still deeply entrenched in keeping the winemaking process alive through the Middle Ages. Wine was a part of Mass. Water was still not safe to drink, so wine was the safest option. It was used in cooking and preserving meat. It was used as medicine to treat illnesses and pain. And wine served a purpose for social activities again.

But as capitalism in the region spread, that created a new type of demand for wine that the church had to keep up with. The citizens were enjoying it more than a means of survival. Wine was back to the pleasurable roots of ancient times. That brought in a much-needed resurgence and respect for wine.

This put a renewed interest in monks honing their craft in winemaking. Monks used their time to experiment with plantings, observing which grapes did better in certain areas. That moved them to divided up pieces of land that were better suited for specific grape varietals based on soil types, weather and overall climate conditions. They also challenged themselves to fix certain issues they encountered while making certain styles of wine.

There was still a lot to figure out. Wine was still being mixed with other ingredients to flavor it, from honey to herbs to spices. And wine didn't have a very long shelf life, so there might have been a lot of wine being consumed that was "past its prime." Under the monk's control, wine flourished with major surpluses that expanded the practice of winemaking from religious purposes into a full-fledged enterprise.

Wine would become so highly demanded and such big business that royal and wealthy households would acquire a steward to order, look after, inventory and maintain their wine and food pantries. Those individuals served in the capacity of an early-day sommelier. They would also look after these items during transportation and shipping.

With new trade routes and partners, wine – along with vine clippings and improved wine practices – were shared with others around the region. That helped to further reignite an interest in wine and honing the craft.

Wine started to increasingly become a luxury item, again.
It was an entree to the finer things in life like fine art, theatre and literature. These things represented the good life, all of which could be achieved by human effort.

: the gateway to the modern wine world

One of the major elements engrained in the DNA of humans since the beginning of time has been the desire to move from place-to-place, exploring the world for new areas, new opportunities and new treasures.

The European explorations of the late 1400s were launched from both necessity and curiosity.

During the Middle Ages there was widespread food shortages and famine in Europe due to a series of devastating weather patterns. Consequently, these areas that were forming into modern-day Europe were looking for either new trade partners or new lands to cultivate themselves.

With the fall of Constantinople in 1453 – the former Christian capital in the East of the Roman Empire – trade routes were cut off between Europe and Asia under the Islamic Ottoman Empire. Therefore, European nations had to find alternative routes to trade and source valuable goods from Asia.

An explorer by the name of Bartolomeu Dias found a solution to the trading route problem from Portugal. He and his men headed west from Portugal, sailing along the West African coast and eventually made it to the tip of South Africa. He is recorded in history as the first European to reach that part of the continent in 1488. In turn, he called it the Cape of Good Hope. That adventure solidified a new route around Africa to deal directly with their Asian trading partners, bypassing blocks set up by the Ottoman Empire.

In Spain, there was the same desire to set up their own trade route to Asia. However, a different route was mapped out by the Italian sailor Christopher Columbus. King Ferdinand II and Queen Isabella of Spain decided to commission Columbus to find a direct route to Asia through an exploratory voyage in 1492.

Columbus never did find a direct route to Asia. Due to his nautical miscalculations, he and his men accidentally stumbled upon the Bahamas region of the Americas.

While there were millions of indigenous people already living throughout the Americas and despite the Vikings having landed there about 500 years before Columbus, he returned to Spain in 1493 reporting to the king and queen about this "new world" he discovered along with its potential for "untapped" treasures.

: the 1500s — the age of exploration

Sparked by news of "New Worlds" by Columbus, exploratory trips increased out Europe, including Columbus who would make three more trips to various parts of the Americas.

Spain and Portugal were quick to satisfy their curiosity of what else was out there to discover. They were eager to expand their kingdoms, their religious beliefs and their wine production capabilities. Eventually colonies started to develop in the new areas they conquered. Colonization also provided them with land in which to plant grapes throughout North and South America.

The planting of vinifera grapes in these areas changed the trajectory of the wine world – officially expanding it across the planet.

Spanish conquistadors quickly settled in the southwest region of North America, now known as Mexico and New Mexico. They were sent to convert the Native American Pueblos to Christianity and try to cultivate a wine region with the "mission" grape varietal around 1521.

In South America, grapes were also planted by Spanish and Portuguese colonizers in places that would later become Argentina, Peru, Brazil and Chile. Spain and Portugal were quickly growing their kingdoms throughout select areas of the Americas and uncovering the winemaking potential of these new wine regions.

Over in Europe, monks were discovering some new styles of winemaking – a wine that included carbonation. In the South of France, in an area called Limoux, papers were written by Benedictine monks about the discovery of a white sparkling wine in 1531. This is the first documented source of sparkling wine, with a secondary fermentation happening in a flask sealed off with a cork stopper. It was called Blanquette de Limoux.

This style of wine would prove to be an intriguing challenge for the monks. They would wrestle with the concept of trying to understand sparkling wine over the next hundred plus years.

Another thing the church would struggle with around this time would be the corruption a group of monks perceived was going on within the Catholic Church. Led by a monk named Martin Luther, shakeups and divisions were starting to occur in the church. It resulted in the movement called The Reformation / Protestant Reformation. This break in the church created to a new group of Christians – Protestants – which also lead to different ways of approaching the Bible and their daily lives from a religious perspective.

One of those shifts was their relationship with wine and alcohol. Many Protestants believed in not partaking in wine at all. The consumption of alcohol was believed to lead to sinful thoughts and actions. If one was to consume alcohol, it was to be in very strict moderation.

By the end of the 16th century, European kingdoms were involved in a series of conquests and conflicts on their own turf. There was a constant struggle for land, power and influence. Those conflicts would spill over to the colonies in the Americas, fighting over who would lay claim to these sections of land throughout in the New World.

: the 1600s — the age of discovery

As enlightenment was washing up on the shores of Europe and flowing throughout parts of the continent, there was a strong undercurrent of turmoil.

It seemed that Europe was having major growing pains and a bit of an identity crisis. It was struggling to find itself as it was developing into the Europe we know today.

That awkward phase forced some Europeans to take action to either make things better or run for their lives. A select group of Europeans, who referred to themselves as Pilgrims, fled Holland and England trying to escape growing turmoil and "unholy" temptations in search for religious freedom in North America.

The pilgrims considered themselves separatists in many ways – mostly severing ties with the English Church – but were still considered citizens of England under the new colonies.

The first colonies were formed in the early 1600s, with colonies setting up along the east coast of North America from Plymouth, Massachusetts to Jamestown, Virginia.

Because of the strict religious and puritanical roots of many of the colonists, the production and consumption of wine wasn't a priority. However, that shifted in 1619 as England enacted the Jamestown Assembly Act in Virginia that required every household to plant ten vines of the vinifera grape species to produce wine as part of the economy.

The eastern coast of North America was already heavily populated with wild, native vines from the Vitis labrusca species. However, the taste of wines made from those grapes were not pleasing to the settlers or to those back in Europe.

This led to some very trying times in the wine world in colonial North America. Not only did they not like the wines made from native grapes, they were also having a very difficult time trying to grow and produce wines from the vinifera species on the eastern seaboard.

They couldn't quite figure out the problem. While England did participate in some winemaking activities, the size and weather conditions of the country didn't quite work in terms of creating their own major wine industry. Only a handful of vineyards and wineries existed in England, so they would rely mostly on wines from nearby regions to satisfy their wine appetites.

France wasn't involved with the English colonies or their winemaking efforts. In fact, the two countries were actually constantly at war with each other. Therefore, France set up their own individual colonies on the east coast of North America, as far north as Canada, Acadia and Maine.

Ultimately, it would take more than 200 years for winemakers on the east coast of North America to figure out why they were having problems with the vinifera vines.

On the west coast, in Mexico, however, the Spanish missionaries were creating a bolstering wine industry. It became so successful that the Spanish King ordered nearly all winemaking efforts there to stop in 1699. The king believed the New World wine production posed a threat to the Spanish colonies' dependence of wine production in Spain. Wine continued to be made in the Mexican regions, but strictly for use by the church.

In the Catholic Church in Europe, monks continued to work hard to study, experiment and make tweaks to their grape growing efforts and wine production. One Benedictine monk and wine master in particular named Dom Perignon was investigating the concept of sparkling wine around 1688 in the Champagne region of northern France.

Wine in the Champagne region initially was reddish in color, made from Pinot Noir and had no bubbles. Residents in Champagne wanted to produce red wines like the ones that became very popular and well respected in Burgundy, France, a little south of Champagne. However, the weather conditions didn't allow the Pinot Noir grapes to ripen as well as they did in Burgundy.

But that was not Champagne's main problem with the wine.

Cellars and underground caves in Champagne were used to store wine from the harvest, which is in the fall, until the wine was needed. There was a mysterious thing going on in the cellar when the weather warmed up slightly in the spring and summer months. Bottles of wines started to explode. That, of course, led to dangerous shards of glass flying around the cellar, hitting other nearby bottles and causing a chain reaction of bursting bottles. It also proved dangerous for anyone working in the cellar at the time of the explosions.

Champagne soon became known as the "Devil's wine." If the bottles were not completely destroyed in the explosions, the surviving bottles would be filled with slightly bubbly wine. That, at the time, was considered to be a major flaw in the wine.

That is when Dom Perignon was tasked with the job of eliminating the problem, or in other words, getting rid of the bubbles.

Perignon worked on the issue but did not come up with solutions on how to solve the bubble problem. However, before his death he made some great advances in winemaking practices within the Champagne region of France. And his work and writing on the subject would prove as inspiration for new Champagne winemakers to come along in the 1700s.

While many have credited him for discovering sparkling wine, most historians refer back to the initial finding in the South of France back in 1531 as its origin – almost 100 years before Dom Perignon was born.

On the other side of the world, in the Southern Hemisphere, the vinifera vines would find a new home.

About 200 years after the Portuguese explorer Bartolomeu Dias sailed around South Africa and coined that area the Cape of Good Hope, the region fell under Dutch rule in 1652. It became an official colony of the Dutch East India Company, a government funded corporation of trading companies. There was a strong demand for wine by the sailors of the trading explorations, so vines were brought over from parts of Europe and planted in the Cape Province in South Africa. This effort would add to the growing number of Southern Hemisphere wine regions growing vinifera grapes – along with South America.

: the 1700s — the age of new beginnings

Beautiful new beginnings and fantastic firsts were taking root in the 1700s. The world seemed poised to come into its own in several different ways – creating new demarcated regions, new tastes in wine, new attempts to spread the wine world and new shifts in power.

As the Catholic Church was becoming more fractured and its power was weakening, the role of winemaking started to transfer over to a négociant system. That system was made up of small growers, land owners and winemakers. They were now making wines under their own names and labels in France sometime starting between 1700 and 1730.

A new thirst for sparkling wine – the style that Dom Perignon was trying to put an end to – spread throughout France with improved methods that enhanced the production value of Champagne during this century.

Much of that interest was sparked by Philippe II, Duke of Orleans and the nephew of Louis XIV. Phillippe II, who began to govern France in 1715 after the death of this uncle, was a big fan of sparkling wine. He often served Champagne at events and dinners he hosted at the Palais-Royal where he set up government operations.

Demand quickly started to increase and expand throughout Paris and among high society. The British also became major converts to this style of wine. To supply this new demand, winemakers in Champagne started to shift their focus from making still red wines to making sparkling white wines.

This would ultimately give rise to the most respected sparkling wine region in the world. As the 1700s progressed, Champagne house after Champagne house stated to form: Ruinart in 1729, Taittinger in 1734, Moët & Chandon in 1743 and Veuve Clicquot in 1772.

From there, the race was on to promote their individual Champagne styles – which were fairly sweet at the time – to royal courts throughout Europe, further spreading the gospel of Champagne.

As select populations in Europe were celebrating life with Champagne, the thirteen colonies on the east coast of North America were celebrating their independence from Britain in 1776. They had won the Revolutionary War, drafted the United States Declaration of Independence and started a brand-new country.

With George Washington as the first president of the United States in 1789 and Thomas Jefferson as the ambassador to France, wine flowed often and freely amongst the politicians and dignitaries. Wine was used for entertaining during dinners at the President's House in Philadelphia and Jefferson served as Washington's unofficial, personal sommelier.

Jefferson had spent some time visiting and studying wine regions in Italy and France. When he returned, he was consumed with a passion for creating a wine culture in the United States as big, if not bigger, than that of France in terms of growing and consuming. His attempts at farming grape varietals from both France and Italy proved unfruitful before and after becoming the third President of the United States.

The United States wasn't the only New World country having issues with the vinifera grape species taking root at this time – at least initially. An effort to expand wine production under the British government took place in either 1787 or 1788 in Australia. Clippings from vines from the Cape of Good Hope in South Africa were taken to Australia to start wine production. Initial attempts failed, but continued efforts and the arrival of new immigrants finally got wine production off the ground in the early 1800s.

As the century was coming to a close and as the Catholic Church was continuing to lose its power in Europe, the Kingdom of France was strengthening. It was continuously trying to establish its borders, acquiring much of the church's wine land holdings.

This would mark the end of church-led wine production for the general public. Winemaking was now in the hands of individual land owners, farmers, winemakers, négociants, families and companies.

: the 1800s — the age of culminations and devastations

The 19th century gave rise to a series of fortunate and unfortunate events.

Thomas Jefferson was now the third president of the United States and living in its new capital in Washington, D.C. While his efforts to foster the cultivation of the vinifera grapes along the eastern seaboard and on his Monticello estate in Virginia were not working, winemaking was being spread by Franciscan monks from Mexico to the San Diego and Sonoma regions of California by 1805.

Vitis vinifera vine planting continued to spread in Northern California with the first vines being planted in the Napa Valley in 1839 by George C. Yount. That area would later be referred to as Yountville in Napa. Then the first commercial winery, Charles Krug, was founded in Napa in 1861.

Success with the wine grape species was also finally happening in Australia. About 33 years after initial attempts to plant vines, the country was now producing and selling wine locally in the 1820s. Then, those clippings from Australia were brought over to New Zealand, establishing the first vineyards in 1836.

During this century, the Champagne style of sparkling wine continued to evolve in France with the help of more technological advances and an ever-changing palate.

With more understanding of what sparked the carbonation in the bottle, Champagne makers began to experiment with the style more. And with Britain being in the midst of the Industrial Revolution, the country was able to produce stronger, thicker wine bottles that were capable of containing the pressure to avoid the spontaneously bursting bottles in the cellar.

Barbe-Nicole Ponsardin Clicquot, better known as Veuve Clicquot (the widow Clicquot), was credited for creating the first rosé champagne in 1818 by blending both red and white wine together – a process that would be adopted and used to make most rosé sparkling wines around the world.

The British would continue to influence how Champagne was being made, by requesting options with lower sugar amounts. The sugar levels of Champagne during the 1800s typically ranged from 110 grams per liter to about 330 grams. The level depended on the market it was servicing. This would make the wine more like a desert sparkling wine.

British fans of Champagne wanted their sparkling wine to have much less sugar, ranging from about 20 to 65 grams. That style would be considered semi-sweet or sweet. That was still much sweeter than the styles more common today. However, those slightly drier styles of sparkling wines inspired Champagne houses and paved the way for Perrier-Jouët to make a Brut style sometime between 1846 and 1854. This new style was initially considered "brutal" on the palate by many people, which gave birth to the Brut name. Brut, of course, eventually grew in popularity and is now the standard dryness level for most sparkling wines.

With the new styles and high demand for Champagne, additional houses started to pop up in this century: Perrier-Jouët in 1811, Bollinger in 1829, Krug in 1843 and Pommery in 1858.

The year 1848 proved to be a standout time that would have major effects on wine that century.

It was that year the Malbec grape made its way from France to Argentina, South America. Typically used as a blending grape in Southwest France, the grape would go on to become the major grape varietal in Argentina about 150 years later.

Mexico had been defeated in 1848 by the United States during the Mexican-American war. That military defeat led Mexico to sign the Treaty of Guadalupe Hidalgo, which forced the country to hand over land to the United States. The areas include states that are now Arizona, California, Colorado, Nevada, New Mexico, Utah and Wyoming.

California was now part of the United States. That year was also the start of the Gold Rush. Many settlers on the east coast crossed the country in their pursuit of wealth in California. Some even took their vines clips from the eastern seaboard. And among many of those new settlers who found wealth through gold, some used it to fund new vineyards and wineries on the west coast.

By mid-century, it seemed like the wine world around the globe was on the path to a very bright and prosperous future. More and more scientific advancements were being made in France and other parts of Europe. Winemakers were getting a better sense of how fermentation actually worked from a scientific perspective.

Over in Bordeaux, France, the quality of wine was excelling to the point that Napoleon III instructed the Chamber of Commerce to develop a classification system to rank the best chateaus based on reputation, trading price and overall quality.

This would be called The Bordeaux Wine Official Classification of 1855. It set in place a high standard of quality for the region to maintain and uphold. Although people question the rationale on how these producers were classified, this move became a major game changer in the wine world. It set the stage for a quality hierarchical system to be implemented in many regions around the world several decades later.

There was so much growth globally in the wine world in these first 50 years. There was so much promise. And there were so many accomplishments made during this time with vinifera grape vines.

Then a small bug called Phylloxera, also known as Phylloxera vastatrix, nearly took that all away. Phylloxera is a microscopic parasite that feeds on the roots and leaves of grapevines. Once that happens, it ultimately injures the roots to the point where they can't allow the needed water and nutrients to circulate throughout the vine.

Remember that issue with growing grapes on the east coast with the Pilgrims all the way to Thomas Jefferson? This was the issue they were facing. These early farmers and grape growers were able to farm the indigenous grape specifies, Vitis labrusca, but the Vitis vinifera clipping from Europe would not work.

They just didn't know the cause of the problem. What they also didn't realize is that when taking clippings of the labrusca vines back to Europe, they were carrying this parasite back with them on the ship around the 1860s.

It didn't take long until the small number of vineyards that existed in Britain started to be destroyed by Phylloxera. The bugs were then unknowingly transported to France. Within a few years, it spread and destroyed about 70 percent of its vineyards by 1870.

The bug then continued to spread to other parts of Europe through travels and then to the west coast of North America. By the end of the century, the bugs ate their way through the Napa Valley region and destroyed about 80 percent of its grape vines.

While a small selection of varietal vines and regions were safe from the pest, it created mass devastation. And it, no doubt, also created a lot of panic. There didn't seem to be cure for Phylloxera.

After some time, a solution was conceived. A Texas entomologist by the name of Thomas V. Munson, in collaboration with Charles

Valentine Riley and J. E. Planchon, developed a solution on how to work around the problem.

Munson realized that the grapes that were native to North America didn't suffer from Phylloxera. Since they were both native to North America, the vines developed a natural resistance over time against the parasite. That is because the roots of the Vitis labrusca vines developed a very sticky sap that repels the pest. That didn't occur with the Vitis vinifera species. Therefore, Munson recommended that the European vines be grafted, or spliced, onto American rootstock and then planted.

After about 16 years, the theory proved to work. It allowed areas devoured by Phylloxera to start the process of replanting. While this did give the wine world a fresh start, the damage was already done. Some grape producing regions were not able to rebound and some grape varietals were lost forever as a result.

: the 1900s — the age of test, trials and triumphs

The thing about history is that it tends to repeat itself. Some things just never change, for better or for worse.

At this point in time, the 20th century, it had been a few thousand years since mankind formed nesting civilizations where they settled, built homes, grew grapes, reared livestock and set up functioning societies.

Some of those ancient desires to continuously expand their communities, empires or governments continued to carry over to the early 1900s. That included the acquisition of more land, the search for new resources, the spreading of political and spiritual beliefs and – unfortunately – the quest to dominate over other people through the use of force and power structures.

One major difference, at this time in history, was that a new, up-and-coming super power country had come into in the picture.

Enter World War I.

For four years, from 1914 to 1918, the "world" powers were in conflict with each other. Britain declared war on Germany in 1914. Then it became the Allied Powers of France, Russia along with Britain – and later the United States – against the Central Powers of Austria-Hungary, the Ottoman Empire, Bulgaria along with Germany.

This also affected, and included, any of the colonies belonging to these major world forces at the time, making it a global conflict. The Central Powers were ultimately defeated in 1918, which had a dynamic impact on the world and the rest of the century.

To suggest that it was a significant war would be a gross understatement. This first world war led to so many residual effects: mass destruction, the loss of millions of lives, changing dynamics on the world stage and a new world economy.

The war sparked an even bigger industrial revolution, igniting a huge economic boom. That was primarily felt in the United States which didn't experience actual combat on its soil. Much of the fighting took place in various parts of Europe, the Middle East and Eastern Europe – including some wine regions within France.

Since there was no destruction to the factories and manufacturing operations within the United States, the country thrived economically and grew in wealth and power. Much of the world started to depend on the U.S., who was making the lion share of manufactured goods.

As a result, the world economy was changing. It was moving into much more of an international economy dominated by the United States.

After the war, the United States entered into a very joyous time. It seemed like everything was available in excess to select populations in bigger cities: money, nightlife, jazz, alcohol and wine.

The wine world seemed to be rebounding nicely from Phylloxera in both the affected old world and new world regions.

Over in Europe and the Middle East, those empires and governments were going through a massive reconstruction period, including having new boundaries created to form new regions. That led to devastating inflation, chaos and uncertainty throughout many of the defeated regions under fresh leadership.

Back in America, the good times didn't last for very long. There was some trouble bubbling up for some time now and not everyone was celebrating with jazz, liquor and wine.

Just when it seemed like the wine world was starting to rebound in the United States, it received another powerful blow. This time, it was dealt by mankind.

A social movement led by evangelical Christian groups, which turned into a political movement, prohibited the consumption of any alcohol in the United States under the 18th Amendment to the United States Constitution. The group had growing concerns for some time about alcoholism and its negative affect on the family unit, the workforce and the overall community.

As a result, many wineries that survived Phylloxera or bounced back from it were forced to shut down. Alcohol was only allowed to be produced for medicinal purposes – prescribed by doctors – and for sacramental purposes. In these situations, some wineries were allowed to stay in operation across the country. However, there was estimated to have been less than 100 wineries in operation in the U.S., out of about 2,500 before Prohibition.

Through illegal sources and some legal loopholes, a large number of U.S. citizens found ways to obtain wine, beer and spirits. But, outside of the United States, the wine world was experiencing small glimmers of hope.

In Europe – particularly Austria – an interesting new farming philosophy was being spread by Austrian philosopher Rudolf Steiner in 1924. Farmers from Austria, Germany and Poland were experiencing deteriorating soils from the use of chemical fertilizers developed during the Industrial Revolution.

Steiner developed a concept called Biodynamic farming. It is a holistic and organic approach to farming which is based on the lunar cycle or the moon's sowing and planting calendar.

I know what you're thinking: What just happened? No worries. You'll learn a little more about this type of farming in the next chapter.

If it sounds strange, you are not alone. It also sounded a bit obscure to the larger farming community for a while. But a several decades later, farmers, grape growers and winemakers around the world would start taking the concept a lot more serious.

Then over in South Africa, Abraham Izak Perold, a professor of Viticulture at Stellenbosch University was making a unique discover of his own. In the early 1900s, South African winemakers were having a difficult time growing the grape varietal Pinot Noir in the Cape wine regions. Pinot Noir was the red wine grape varietal made famous in the region of Burgundy, France.

In 1925, Perold created a new grape varietal at the university that became unique to the region. By cross breeding two Vitis vinifera grape varietals – Pinot Noir and Hermitage (called Cinsaut in France) – he was able create a new grape varietal called Pinotage (**pinot** + hermit**age**). The effort was successful in both planting and growing, officially making Pinotage the primary red grape of South Africa.

Back in the states, Prohibition lasted for 13 years and ended in 1933. But when the law was repealed, the United States – and the world dominated by its economy – was three years into an international Great Depression. It hit many of the major cities around the world hard, but also had a huge impact on farming and agriculture.

By the end of the 1930s, several of the winemaking countries were trying to be revived where the Great Depression hit especially hard – like Italy and Germany. The U.S. wine industry was also working on recovering both from the economic downturn and the effects on the country's vineyards during Prohibition.

Yet again, some things just never change. History was about to repeat itself very quickly in the form of World War II.

The world was at war with itself again – this time – starting when Germany invaded Poland in 1939. This pitted the Allied Powers of Britain, United States, the Soviet Union and France against the Axis Powers of Germany, Italy and Japan. As with the first world war, it effected and involved any colonies related to their parent countries. Therefore, making it another global conflict. The war ended in 1945 with Axis Powers surrendering.

This second war would have an equally powerful impact on the world as the first one. During the war, there was a crucial need for all-hands on deck this time. That called for factories to reopen, new technologies to be developed, increased food production and an increasingly larger labor force. That put women in many new working roles they hadn't had access to before.

World War II also helped to further establish the United States as a major super power in the world. The war continued to redefine borders in Europe and some of its former colonies. And it pushed Europe toward the rebuilding and reconstruction process, bringing them under a more unified political and global identity.

Winemaking had continued throughout the world during the second war when and where it was possible. Once World War II was over, special attention was back on wine again throughout the world with an increased interest in new wine practices and technologies.

After the war, a lot of attention was focused on Napa Valley to help jump start the wine industry in California. That started by reestablishing many of the vineyards like Beaulieu Vineyards in 1938.

American life was now different moving into the 1950s. Or, at least, that was the hope and expectation of many citizens. There was a new focus on convenience and experimenting with ways to make life easier, more efficient and more abundant.

The focus was on better days: fun, exploration, discovery. People wanted something memorable and lively to yank them out of the melancholy memories and dull days of the economic depression and the second world war.

This was true in other nations around the world as well. Areas from the Soviet Union, East Asia and Western Europe all experienced unprecedented and prolonged economic growth – most of which lasted for about 30 years.

There was a real financial boom felt from Britain to Italy to France to the United States, ignited by new ideologies about how many problems in society could be remedied by science and technology.

Consequently, technology was everywhere. There were advances in stainless steel, refrigeration, television, film, home appliances, frozen dinners, fast food restaurants and mass-produced items.

The chemical advances learned during both world wars were further applied to the agriculture industry worldwide. This new approach, referred to as "Intensive farming" or "Industrial agriculture," placed a new and expansive scope on higher yielding crops. This new approach was in contrast to traditional farming.

The Industrial agriculture revolution moving into the 1950s was made possible by a more aggressive management of the soil, water, weeds and pests. There was increased use of chemical fertilizers, herbicides and pesticides and new mechanized farming with the use of tractors and other equipment.

Science and technology were also applied to the global wine world. The techniques in grape farming and wine making significantly evolved during this time.

The science of wine was being understood more, so developments in refrigeration, stainless steel tanks, smaller oak barrels, pressurized tanks, machine harvesters, Stelvin screw caps, bag-in-a-box technology, sorting tables, electric presses and the understanding of wine faults would be developed over the next 50 years.

Winemakers in France, Italy and Germany were producing wine for a more discerning national and international wine market. These winemakers were focused on tradition and tried-and-true techniques, while still working to improve their wines. Therefore, France and Italy became "gold standards" for wines around the world.

Other winemakers, established or up-and-coming, were interested in learning those techniques, so many worked and studied grape growing and winemaking in certain European regions. Then they would take those lessons with them to their respective winemaking regions in the United States, South America and Australia and apply that knowledge to their wines. And, in some cases, winemakers from France and Italy were hired by New World companies to lead their brand's production.

This new look at wine also spilled over to the culinary world. There were a number of fine dining establishments making a name for themselves, touting the best of the best food and wine options from France and Italy.

That also inspired the upgrading of home entertaining. With the help of Julia Childs' cookbooks and television shows promoting French cooking at home, Europe was looked at as the place for refinement and sophistication.

From this stretch of time, from the 1950s to the 1970s, American winemakers were getting really serious about producing higher quality wines from the vinifera species. This was the case in the Finger Lakes region of New York State, where the focus on quality was a main priority, and also in Napa Valley, California, where the dreams of Thomas Jefferson were to be realized after almost 200 years.

In 1976, Jefferson's dream was fulfilled. An English wine merchant and wine educator in Paris by the name of Steven Spurrier decided to try a little experiment and pit French wines against American wines in a blind-tasting challenge.

That challenge was called "The Paris Wine Tasting of 1976." American wines beat out French wines in both the white and the red categories – with a Cabernet Sauvignon by Stag's Leap Wine Cellars taking top honors for the red wine and a Chardonnay from Chateau Montelena winning in the white wine category.

America was already considered to be a strong military and economic force. Now it was being known for making some of the best wine in the world, which really turned the wine world on its head.

Napa Valley producers saw this as a unique advantage to step up their game across the region. And that sentiment had a ripple effect around the world. The wine world shifted tremendously as the global competition to be a known and respected wine producing country increased.

Even China, a country that had been exposed to the vinifera wines thousands of years ago, started to import French wine and encourage French winemakers to plant vineyards there in the late 1970s.

Procuring fine wine was the new goal.

A new rating system was launched in 1978 by an American wine critic named Robert Parker that help consumers make some sense of what might or might not be considered "quality wine." The newsletter, turned magazine, gave people a new way to categorize wine in the form of points from 50 to 100.

Unfortunately, not all regions were going to enjoy riding this new wave of popularity of fine wine by the mid 1980s and 1990s.

The wine making region of Austria would experience a huge scandal in 1985. Several winemakers in Austria were adding an artificial sweetener to their wine called diethylene glycol. This liquid is commonly used in the making of things like antifreeze, dye, brake fluid and cigarettes. The chemical is considered toxic and is not allowed in food or drugs. After media reports connected the sweetener being added to the wine, it nearly ended the Austrian wine market.

Then in South Africa, winemakers were blocked from entering the international market. The region had been making wine since the 1600s and developed some very famous fans in France and other parts of Europe.

Starting in 1948, a governmental system of racial segregation called apartheid was implemented to keep Black Africans, "colored" and Asian South Africans legally and financially separated from the minority white population of Afrikaners. Due to that policy, many other countries boycotted their wines and wine was only sold within Africa. Apartheid was later dismantled in 1994, which allowed South Africa to slowly enter the global wine market.

With a highly competitive industry, wine started to be sold at higher and higher price points globally. Financial excesses within certain industries and pockets of wealth helped lead to wine pricing extremes.

In many cases, wine was elevated to a luxury item and became a status symbol around the world. Select wines produced from high-profile locations and by well-respected winemakers were being sold for thousands and thousands of dollars per bottle.

On the other hand, there was, of course, an even larger market for affordable options.

Many of the value brands became more sophisticated in their branding, packaging and marketing efforts to speak to savvy wine drinkers and potential new audiences.

Due to improving computer technology, 1994 marked the beginning stages of an online wine store that would eventually become Wine.com.

In addition to a century that had a lot going on globally, it was also the end of the century and the end of the millennium. This marked a time of great excitement, anxiety and uncertainty for the countdown to the new century.

: the 2000s — the age of new beginnings and distinct wine worlds

The world was celebrating.

People were partying like it was 1999, because, in fact, it actually was. That was one reason for celebrating. Some people had been waiting for that night for 17 years since Prince first sang about it in 1982.

Another cause for celebration was because it was officially turning into the year 2000. It was not only the beginning of a new century; it was the beginning of a new millennium.

Lastly, people were just relieved that the world did not implode technologically as predicted by the Y2K theory.

Do you remember that?

It was the scare that as we entered the 21st century, the computers would stop working. And not a personal computer problem, but a world-wide shut down that would create a complete breakdown of society – banking systems, electricity, water services, the media and anything you would consider necessary for a functioning global society.

Think the collapse of the Roman Empire times ten thousand.

But ultimately, the world was safe from a global apocalypse by way of computers. Life continued on and people were happy.

Unfortunately, an unexpected disaster would come along one year later, on September 11, 2001. The World Trade Center's Twin Towers were attacked along with the Pentagon in Arlington County, Virginia by an Islamic terrorist group called al-Qaeda.

This attack on U.S. soil put Americans, and the world at large, in a devastating place emotionally, spiritually and financially.

While America was going through a slight recession in 2001, the attack pushed the country and the world into a deeper recession. Some of the effects still linger today.

Unemployment was on the rise. The U.S. was involved in the "War on Terror." And the cost of the attacks racked up quite the bill. A 2011 report from "The New York Times" put the total costs of the attacks to the U.S. at $3.3 trillion.

Americans – and the world – needed a drink, badly. But now, deep into a crippling recession, consumer behaviors were changing when it came to wine purchases.

The wine business is often referred to as a "recession proof" industry. That basically means that when times are financially rough, people are still going to spend money on wine to help them cope.

However, with money being an issue for the majority of people – and the future being uncertain – wine consumers were starting to look at more value-based options opposed to the moderately priced or high-end options from the previous two decades.

This became a major game changer in the wine industry.

At this time, ideal wine prices began to fall to $15 and under for a lot of wine lovers and even $10 and under for many other consumers. Consumers were looking for value, quantity and – in a lot of cases – great quality. Restaurants, wine shops and grocery stores had to quickly adapt to meet the new needs of their customers while also trying to stay afloat until things turned around.

Competition increased for that limited dollar. Some wineries shut down. Some vineyards went into foreclosure, bankruptcies and/or consolidations. While wine was still being produced and consumed, the hardship of the recession was felt by everyone involved.

While the recession continued to drag on, there was a new interest in both value wines and the higher quality wines that might be purchased for special occasions or celebrations.

Winemakers moved to capture both those markets with their offerings. Some going exclusively for the production of value wines costings less than $10, with some options launching in 2002 costing as little as $1.99 per bottle.

Other companies started to offer a mix of wine options at various prices: high-, medium- and some low-priced options under their brand umbrellas. And, of course, the luxury market was still producing their wine for a different, more affluent consumer market.

This pattern of diversity of price points and wine offerings continued, and was potentially expanded on, when the financial crisis of 2008 occurred.

Do you remember that one? It was linked to the "subprime mortgage crisis." Lasting for just more than one year, it was called the "Great Recession" and the worst economic period since the Great Depression.

With the economy being so up-and-down since the beginning of the new century, it brought forth a distinct shift in styles of wine that people would spend their money on purchasing.

With a big focus on value, many of the "under-the-radar" wine regions outside of Europe started to get some major attention. New producers and brands particularly in regions like Central Coast California, Washington State, South America, Australia, New Zealand and South Africa were looking for ways to cut through the clutter and find ways to stand out amongst the competition.

Shiraz from Australia became a big hit and one of the first major trendy wines outside of the major regions of Italy, France, Spain and Napa Valley, California. The wine was big, juicy, affordable and easy to pronounce, so it became very popular from the 1990s to the early 2000s.

Then, right on its tail, growing in popularity from Argentina was the grape Malbec. Coming from another New World wine region outside of Europe, it appeared to have outpaced Shiraz in the race for global popularity – particularly in the United States. It is also very affordable, easy to pronounce and featured crowd-friendly flavors.

Besides economic factors and wine trends, pop culture was also shaping consumer behavior. The 2004 film, "Sideways," starring Paul Giamatti as Miles and Thomas Haden Church as Jack was a road trip dramedy that followed the two main characters on a few days of wine tasting in California's wine country. The movie sparked a renewed interest in visiting vineyards in California and introduced many viewers to "wine speak" and wine tasting terminology.

The movie was also unintentionally responsible for a change in the types of grapes many of these producers grew. After the movie's successful release, less attention was put on the grape varietal Merlot in California.

Based on some very specific dialogue in the movie, more attention was being channeled into growing Pinot Noir. Consequently, Pinot Noir increased substantially in both production and consumption within the U.S. market and Merlot became a major afterthought to many consumers. This has been referred to the "Sideways Effect" by a study done by economic professors.

Other major trends in wine popped up during this time.

White wine started to receive some major attention. Sweet and Off-Dry (a kiss of sweetness to very sweet) wines like Riesling and Moscato started to grow more and more in popularity. Then crisp, refreshing Sauvignon Blanc wines from New Zealand became a go-to white wine for a large number of white wine drinkers.

There was also growing distastes for oaky, butter Chardonnay wines, primarily from California. For several different reasons, consumers were leaning more toward a Chardonnay with little or no oak influence. That is, if they were going to opt for a Chardonnay at all. This classic grape, like Merlot, was quickly falling out of favor for many consumers.

One trend that virtually exploded was the popularity of rosé wines. Drinking rosé these days is the epitome of pop culture. After a while, the wine category even had its own catch phrase and hashtag: #roseallday. These wines, pale to deep pink in color made from red-skin grapes, serve as alternatives to red and white wine in the warmer months. And for most consumers, the dryer the rosé – the less sweet it tastes – the better.

When it comes to trends, however, you have to take the good with the bad.

Society was finally rebounding from the Great Recession of 2008 that touched most people's lives. The economy was growing. Jobs were being created. Trade efforts were increasing. And people were feeling safer about making certain splurges in terms of wine, food and travel.

There are, of course, small pockets of communities that are fairly immune from these types of financial difficulties. At least, they didn't feel the financial pinch of the times, like most people did.

But with more money comes more problems.

The unfortunate trend of counterfeit wine remerged in the early 2000s. It's actually a very old form of deception dating as far back as the Ancient Roman Empire. The counterfeiting and label fraud that occurred during this era, played off the status and prestige that some wines started to carry in the 1980s.

Many consumers were hoodwinked into thinking that wines they spent several pretty pennies on were, in fact, rare treasures in the wine world. Those cases have been documented in a book called "The Billionaire's Vinegar" and in the documentaries "Sour Grapes" and "Red Obsession."

These practices go back thousands of years. In addition to the Roman period, counterfeiting was especially prominent during the Middle Ages when wine was stored and transported in barrels. Merchants and sailors would sometimes mix wine together from different barrels or swap out barrels with inferior wines.

In the late 1700s, Thomas Jefferson was warned about these types of practices in Bordeaux, France. Whenever he ordered wine from that region, he asked that it be bottled at the winery before setting sail to the United States.

Ironically, Bill Koch, an American businessman and billionaire purchased four bottles from a wine collector named Hardy Rodenstock that were considered to be rare bottles that Thomas Jefferson owned. The bottles turned out to be fake. Koch would also fall victim to another fake wine scheme involving Rudy Kurniawan years later, finally reaching a settlement of $3 million in 2014 for his losses.

There were also other groups of wine lovers becoming more discerning with what they were purchasing and consuming.

While getting a great bargain is the general philosophy in terms of shopping, a growing number of consumers were setting their eyes toward smaller, lesser known winemakers and regions. Consumers were starting to put more trust in the craftsmanship of smaller producers. These grape growers and winemakers were perceived to take a more personal, hands-on approach to their wines. Being a small producer soon became a unique selling point.

As an offshoot of seeking out smaller winemakers and producers, consumers were also looking into their farming practices. Winemakers and wine consumers were both becoming increasingly knowledgeable about the negative effects of the chemical fertilizers, pesticides and herbicides that were pushed onto the agricultural world after World War II.

Consumers were becoming more and more suspicious of what they were putting in their bodies since the 1960s. But these days, they have a lot more questions and concerns. Therefore, they want to know if a producer – large or small – is practicing industrial or more natural farming.

It wasn't long before a new group of wine consumers started to emerge. These consumers were interested in grape growers and winemakers that use natural, organic styles of farming. They want to know that they can feel comfortable and safe about the wine choices they are making.

These consumers were fortunate enough to find a growing number of grape growers and winemakers from around the world who were on the same page.

Winemakers, also, started to grow concerned about the health of their land, vines, grapes and themselves. To help solve some of the issues caused by chemical erosion to their soils, these winemakers started to put a renewed interest in old-school, natural agriculture. They want to create something that uniquely speaks to their region, while thinking responsibly about the consumer and the Earth.

While there are a large number of wine lovers who long for tried and true practices, there will always be a growing number of wine lovers who thirst for innovation.

Creativity and technology, therefore, still tend to be leading forces in the wine world. From a packaging perspective, wine companies are looking at the changing trends in consumer behavior and developing new product packaging that best fits their lifestyles.

Box wines, for example, have come back into the spotlight. For a long period of time, this type of wine format was thought of as unsophisticated. The packaging was considered "tacky" and full of cheap, bulk wine. Now wine in boxes – whether they come in a small format tetra pack or the 3-liter bag-in-a-box format – mirror the same quality found in traditional glass bottles. And the box wines offer a value to the customers in terms of volume and how long the wine can last.

Then some wine companies have had success offering wine in cans. The cans range from a 187ML (single serving) to 375ML (half bottle) size. This style of wine, like the smaller format boxes, focus on easily transporting wine from one destination to the next. The concept is to take your wine on the go: to the park, to the beach, to a weekend getaway. It also provides the convenience of enjoying a couple glasses of wine without opening up an entire bottle that might not get fully consumed.

That's never a problem in my house. But for some households, I understand finishing an entire bottle of wine can be an issue.

And winemakers – large and small – are thinking about the environmental costs of doing business. They are keeping a close eye on their corporate carbon footprint.

Wine bottles are heavy. Empty ones, not so much. But when full, they typically weigh about 2.5 to 3 pounds per bottle. The production and transportation of these bottles creates a significant carbon footprint. Consumers severely concerned with the environment love the idea of lighter packaging that helps to reduce those carbon dioxide emissions. That includes the box wines, can wines and new plastic bottles.

Yes, plastic bottles. Wine is now being placed in light-weight, plant-based PET (polyethylene terephthalate) recyclable bottles. These bottles can weigh less than 2 pounds and are considered to not contain BPA (Bisphenol-A) – a chemical known to leach into liquid stored within the containers.

That's just one side of the innovation coin.

In terms of our "Information" and "Internet Age," many wine consumers have been utilizing various web-based technologies to help them with their wine purchases. Whether it's checking the going price on a wine, keep track of favorite wines or shopping for wine to be delivered to a home or office, there seems to be an APP for it.

: the present

Wine has no doubt been through a series of changes. The concept of "wine" leading up to this current moment in our lives has progressed so much.

All of those developments have been absolutely altered, shaped and evolved by time and by so many civilizations. Understanding wine's long, sorted history can help us develop a deeper respect and appreciation for this juice.

That understanding should also lead us to this realization: Wine will continue to evolve.

History indicates that very clearly. We have to remember that everything must change, at some point. Important regions might change. Important grape growing practices might change. Important grape varietals might change. Important environmental factors might change.

Wine will continue to establish new cultural consumption habits. It will continue to gain new fans. And it will continue to offer something for everyone – all around the world – from the strict traditionalists to the experimental early adapters.

This is because wine is civilization's beverage. It belongs to all of us.

Therefore, we have to evolve along with it. If we don't decide to change with wine, there is a big chance, we will get left behind and become estranged from our very own wine world.

CHAPTER TWO: **the place**

Some things in life are just kismet. Written in the stars. Meant to be.

Like when Harry met Sally. Bert and Ernie! Ebony and Ivory. It is said that they live together in perfect harmony.

These thoughts might come off as the trappings of some "happily ever after" fairytale or a Gary Marshall production. But whether wine is like a fairytale come true or, indeed, a gift from the Gods, there is something magical when all the stars align, and particular elements come together to create something special in a very specific place.

In that sense, the wine world is like real estate. It's all about location, location, location.

Well, maybe not *all* about location. But it's a big part of it.

Location gives wine depth. Location gives wine personality. Location gives wine character. Location gives wine an identity.

When the right grapes, the right location and the right people intersect, it's truly a match made in heaven.

There is a symbiotic relationship that is formed between the land, the vine and mankind. When individuals fall head-over-heels in love with the region, the land, the soil, the vines and the grapes, there is unbreakable bond, mutual dependency and natural balance that is achieved.

These all become driving factors that shape our “wine world”.

: the wine world

Having a general understanding about how our planet works is important to learning about wine. This luscious, dynamic and gorgeous Blue planet of ours gives us the atmosphere we need to create and sustain life as we know it.

If you are as far removed as I am from those high school science class days, no worries. This chapter will drop little details along the way to help refresh your memory.

Mother Earth provides the oxygen, water, minerals, nutrients, food, ecosystem and energy that all living things need – the nearly 30 million different types of plants and animals, and of course, humans – to stay alive and survive during our designated life spans.

Earth also gives us a distinct section of her in which we can successfully cultivate the grapes that eventually turn into the wines we consume.

These specific places fall between the 30th and 50th parallel in latitude in both the Northern and Southern Hemispheres of the world, divided by the Earth’s equator.

While these are ideal regions in which to properly grow and harvest grapes, not all regions have suitable conditions for the wine grape vines.

There are exceptions to this. You'll quickly find out in this book that there are several "exceptions to the rule" in the wine world. For example, some wines are made in other places that fall outside of these specific ranges, like Brazil and the Canary Islands.

In the case of the Northern Hemisphere, from the top-down – the North Pole toward the Equator – the latitude for grape growing goes from 50 degrees latitude in the north to 30 degrees in the south.

Wine regions that populate this half of the winemaking world include Austria, France, Italy, Spain, Germany, Portugal, North America, Canada, North Africa, Eastern Europe, The Middle East, China, Japan and a portion of India.

The 50th degree latitude represents the coldest conditions in which grapes can be properly grown. Further north of that top range, it gets too cold to suitably grow and cultivate grapes. And the 30th parallel represents the southernmost extremities and warmest areas where these grapes can effectively grow to be turned into wine. Too far below that and it gets too hot to cultivate grapes for winemaking.

To put things into perspective, the Champagne, France, wine region is located at the 48th degree in latitude, closer to the North Pole. Then there is the Mount Edna region in Sicily, Italy. That area is located at the 37th degree latitude. And just below that 30th degree parallel falls the Canary Islands at 28 degrees, a bit closer to the Equator.

Then the wine world's life got turned upside down when the vinifera grapes were planted in the Southern Hemisphere. Dig the "Fresh Prince of Bel-Air" reference?

In the Southern Hemisphere, the 30th parallel is in the north, closer to the Equator, where temperatures are warmer. That puts the 50th parallel in the south, closer to Antarctica and the South Pole where temperatures are colder.

For a better perspective in terms of wine regions here, Cape Town, South Africa is at about 34 degrees latitude, closer to the Equator in the north while Patagonia, Argentina, lands close to 42 degrees closer to the South Pole.

Whew! Wow, I know. Take a pause if you need one. Those were a lot of details. The "less is more" approach is a bit challenging to apply in this chapter.

In essence, just remember that these boundaries mark the extremities in which the vinifera grape species can grow properly in the wine world – not too hot, not too cold, but just right!

While the wine world is physically divided into two hemispheres on the planet, it is also divided into regions with different approaches to wine based on history and philosophies.

Those are called the "Old World" and the "New World" wine regions.

Some winemaking regions have been making vinifera wines for thousands of years as we learned in CHAPTER ONE. These regions are called the "Old World" as referred to by historians.

Then there are regions that have been making vinifera wines for a few hundred years in places outside of the traditional "Old World" regions. These are places that have been colonized and/or were ruled by old-world regimes and, as a result, are referred to as "New World" regions.

All of these regions – Northern Hemisphere, Southern Hemisphere, Old World, New World – make up what is called "the wine world."

Let's take a closer look into these distinctive worlds and styles to see how they differ.

: the old world versus the new world

Knowing the terms "Old World" and "New World" can truly help put the overall "wine world" into perspective.

Getting familiar with the uniquely different styles those two worlds create will help you develop a strong sense of why "the place" is so important in the development of wine grapes.

There are several traits that distinguish the two regions and styles from one another.

As a wine educator, one of my favorites wines classes to teach is "Old World versus New World." In this class, I do my best to conduct an apples-to-apples, side-by-side comparison of the same wine grapes from the different regions for students to sample.

For example, I'll feature a Sauvignon Blanc from France next to a Sauvignon Blanc from New Zealand. And then a Pinot Noir/Pinot Nero from Italy against a Pinot Noir from Oregon in the United States.

To me, that is the best way to physically experience how they differ, by tasting them side-by-side. It's really quite fascinating when you get to taste and smell the differences.

However, sometimes confusion can set in when trying to distinguish the two if there is a perceived overlap of these styles: like, if they look, smell and taste too similar. It's almost feels like the two styles have blended into one style, which can muddy the water a bit.

I always say that wine is not always so black and white, there is a huge grey area. Or if I want to be cheesy and corny – which is sometimes very fun for me as I'm originally from Wisconsin – I'll say wine is not always distinctively white or red, there is a huge rosé area in wine.

I warned you: cheesy and corny.

The large overlapping of styles has perpetuated confusion due to technology, the globalization of wine and crafting wine for particular consumers and wine critic's tastes.

This section is intended to help you understand the particular ways these two regions differ. Then later, I will showcase how the overlap sometimes happens.

: the old world

"Old World" is a term that includes geographic regions spanning from the Mediterranean, North Africa, Europe, The Middle East and Asia.

That mirrors the wine world as well, with a special emphasis on a number of prominent wine regions like Austria, Italy, France, Germany, Greece, Portugal and Spain.

Winemakers in these regions have been making wines from the vinifera grape species for about 2,000 to 4,000-plus years. Over that time, these professionals had to overcome extreme hurdles: climate change, wars, expanding empires, collapsing empires, challenging landscapes, obsolete technologies, cultural shifts and the ever-evolving wine palate.

A unique and powerful bond between mankind and the land has been formed, as a result. The numerous challenges paved the way for some brilliant agricultural discoveries. These farmers learned how to identify the best ways to nurture and care for grapes in specific regions – practices that were shaped by time and trial and error.

And those learnings – coupled with a deep respect for the land – helped instill a strong sense of place. It became the desire and mission of these winemakers to showcase the inherent personality of the land through each wine that was made. Those developments have become well respected and steeped in tradition for centuries.

To protect these traditions and techniques, certain safeguards have been put in place by many of these Old-World countries. Strict governmental regulations help preserve the heritage of winemaking and protect the unique styles from appellation-to-appellation, region-to-region and country-to-country.

These systems of regulations set out to outline distinctive appellations or sections of lands for grape growing. Then the policies can ban the use of irrigation, select approved grapes that are allowed to carry a particular region's name, limit the amount of wines (overall yield) that can be produced every year, put limits on the alcohol contents and standardize aging requirements on certain wines.

That's just to name a few.

The system also creates a hierarchy of wines in each country, ranging from the top, high-quality vineyards which are heavily regulated to other less regulated regions that produce rustic "every day" wines of good quality.

The major theme in Old World wines is that they are associated with bringing out the best expression of the wine's sense of place. Wine drinkers should be able to taste the grapes as well as the land's environment.

This is why a vast majority of wines that are made in Old World regions are named after the specific place in which the grapes grow: Burgundy, Bordeaux, Barolo, Chianti, Champagne, Provence, Rioja. These are all wines named after their regions. Those are not the names of the grapes found in these wines. These are places.

That's an important distinction to make because in the Old-World philosophy, the land is most important aspect in winemaking. The grapes are an essentially a product of the land or place.

Besides the shared experiences of knowing the land deeply and protecting their various winemaking traditions, it also just so happens that these Old-World regions all geographically fall within the Northern Hemisphere of the world. All those similarities help to give these wines a comparable style that sets them apart from New World wines.

That commonality is greatly shaped by climate. Many of these wine regions fall into cool-climate areas, so the grapes aren't always fully ripe when picked from the vines.

That climate creates wines that are higher in acidity, lower in alcohol and mineral driven. More specific details about acidity, alcohol and minerality are coming up in Chapter THREE: **the elements of wine**.

These differences do not speak to all the wines within this Old-World style, however. There are numerous "exceptions to the rule" like Spain and Portugal. Attention must also be paid to the climatic differences within these varied regions within a country and also from country-to-country.

As we know, within these 30th and 50th parallel boundaries in the Northern Hemisphere we have ranges in climate that showcase cooler temperatures in the north and warmer regions in the south.

While the grapes in a cooler climate region like Champagne, France, might not fully ripen before being picked, the grapes found in the Southern Italian region of Sicily will experience a warmer growing season. These grapes should have no problem reaching their peak ripeness levels.

Those differences will then affect the acidity and alcohol levels, and potentially have an impact on how the mineral structure stands out in the wine.

But with the strong sense of place, traditional farming practices, protected winemaking techniques and similar climates, all these circumstances culminate into an "Old World" style of wine.

: the new world

New World wine regions, on the other hand, are much different.

These are areas spread around the world, outside of the Mediterranean region, that had been inhabited by millions of indigenous people prior to the 1500s.

These cultures had been living independently of the Old-World cultures and practices for thousands of years. Some regions even had their own specific grape species and wine cultures.

That was until exploratory trips increased out of Old-World regions which was sparked by news of "New Worlds" across the Atlantic Ocean by an Italian sailor named Christopher Columbus.

He and his crew accidentally stumbled upon the "Americas" – more specifically the Bahamas region – on an expedition sponsored by Spain to find a direct sailing path west from Europe to Asia in 1492.

The Vikings, led by Leif Erikson, had actually made contact with these areas and the indigenous people about 500 years before Columbus. But Columbus was the one who made enough noise about these places to get the Western world curious about new lands, new luxuries and potential gold and silver reserves.

He made the news go viral, in a sense.

The intel of these areas and their potential spread fast, bringing visitors from countries like England, France, Portugal and Spain to eventually conquer and settle in areas that were previously unknown to them.

The hope was for these countries to seek untapped treasures, expand their kingdom's territories and spread their philosophies, religions and their ideal way of living to these seemingly uncharted lands.

It was also a way for them to create an extension of their wine cultures to these places, cultivating the land to plant their grape species to increase their wine operations and wine trade opportunities.

These missions took place between the 1500s and mid 1800s. This eventually resulted in new wine regions that have now been making vinifera wines for nearly 200 to 400 years, hence the name "New World" wine regions.

Unlike the Old World, the New World wine regions are spread out across the world between both hemispheres. In the Northern Hemisphere, the regions include the United States (California, Washington State, New York, Oregon), Canada, China and Japan.

Then on the flip side in the Southern Hemisphere, the primary winemaking countries are South America (Chile, Argentina, Brazil, Uruguay), South Africa, Australia and New Zealand.

Though the regions are disconnected physically around the world, they are connected in spirit. That spirit is fiercely attached to innovation, experimentation and the freedom to express the grapes and areas how a winemaker best sees fit. These wine regions don't adhere to the same wine laws and regulations found in Old World countries.

However, nature is nature. Since that can't be escaped in terms of harvesting grapes, New World wines are still products of their environment.

In terms of climate, it appears that many of the New World wine regions are a bit warmer than those of the Old World. Since that's the case, the wines from these regions tend to be a little higher in alcohol due to the ripeness of the grapes and have a softer acidity. With the riper grapes, these wines will tend to feature more fruit notes and more subtle mineral notes.

Remember, these are "blanket statements" that cover off on the larger, general "New World" style. These traits don't fully represent every single region in the New World.

With a lack of rules and regulation, these winemakers are able to find new unique regions for grape varietals. They have also been known to sometimes create new wine regions that were not previously suitable for grape growing.

Where irrigation is banned in the Old World, it is celebrated in the New World. Winemakers can also grow any grapes wherever they want. The wines don't have to speak to a protected place or region, although – again – the land is still very important to New World winemakers.

While the land is important, more emphasis is put on the grape varietal, the winemaker and production styles. Technologies used in New World wine production allow the winemakers to blend, ferment and age the wines in ways that add to the intensity of the wine. That gives these winemakers much more creative freedom and control over the final product.

It's quite a different world and philosophy from where Old-World wines come from.

Then there are the times when those two distinctive worlds collide. When that happens, it can prove difficult to tell the difference between the two wine worlds when tasted side-by-side.

: the overlapping worlds

If the two worlds start to seem similar to each other, part of that reason is because of their sometimes-mutual infinity with each other.

There is no doubt that the art, craftsmanship and specific style of wines from the major Old-World regions have had a powerful impact on the New World wine regions – either directly or indirectly.

A lot of times, these newer regions look to the classic ones as an example of how wine should be made. Therefore, some winemakers attempt to make a New-World wine in an Old-World style.

You'll see this happen a lot when New World wine regions are influenced by particular regional characteristics that happen to closely link up with those of an Old-World region.

When that happens, both regions will share some of the environmental conditions that allow the grapes to develop in a similar fashion. That can be said of the Pinot Noir wines that come out of Willamette Valley, Oregon, in the United States, and the Pinot Noir wines that are produced in Burgundy, France.

In terms of latitude, Willamette Valley falls in at about 45 degrees. That climate is on the cooler spectrum when it comes to producing wine, knowing that 50 degrees is the top range. This cooler environment gives the grapes certain characteristics. The grapes from this region can then be pretty similar in makeup to the grapes from Burgundy, France, because this region falls in at 47 degrees latitude.

With the climate being very similar, a Willamette Valley winemaker can produce something that looks, smells and tastes pretty comparable to the Pinot Noir out of Burgundy, especially they use some of the Old-World techniques of winemaking.

It would be similar, but not exactly the same. There will always be something very distinctive about the sense of place in each glass. Still, the characteristics of the wine are similar enough to blur the lines between the two world's styles.

Conversely, the Old World has been inspired by the New World as well. Since the New World wine regions have much more freedom of expression and flexibility to incorporate innovation into their production process, some Old-World regions might borrow some of those techniques to give their wines a bit of a New-World spin.

This is the case for many Spanish red wines. Most winemakers in the Old World will opt to use some type of European oak barrels to age and/or ferment their wines. We'll get into this topic a little more in the next chapter. But Spain will sometimes opt to use American oak barrels for many of their red wines. This helps set Spain apart from other Old-World wine regions and can, as a result, give these reds the look, smell and taste of the New World.

So, it can go both ways.

With the globalization of wine these days, similar traits in climate, winemaking styles, techniques and a number of other factors can allow both styles to bleed into each other. However, getting to know the unique differences between the two worlds can actually allow you to clearly pick out a Willamette Valley Pinot Noir apart from a Burgundian Pinot Noir in a tasting exercise.

That, in a nut shell, is the Old World versus the New World.

We've covered quite a lot of ground so far when it comes to a wine's place: Northern Hemisphere, Southern Hemisphere, Old World and New World.

But as alluded to at the beginning of the chapter, it's all about the location. And most importantly, the specific details of each place: the environment, the aspect, the climate, the water, the soil types, the overall "terrior."

Let's start digging into that!

: the terrior

"Terrior is everything."

That's what grape growers say. That's what winemakers say. That's what wine experts say. That's what history says.

It's all about the terrior.

So, what is terrior?

Terrior is a French word that translates into the word "soil" in English. This short word, however, extends so much further than just the soil of a parcel of land. As a result, the meaning can often get lost in translation.

The concept of terrior basically encapsulates the overall environment in which the grapes are grown on the vines. That takes into account the soil – the mineral makeup of a piece of land – as well as the climate, the weather, sun exposure, vineyard conditions, proximities to bodies of water, altitude and other variables.

The general concept, which also extends to other food and beverage products, was said to have been used by winemakers in Ancient Greece, Egypt and Rome but officially developed by Benedictine and Cistercian monks during the 11th century in France.

These monks divided up pieces of land that were better suited for specific grape varietals based on vineyard conditions. That practice of opting to plant specific grapes best for a land's distinct terrior can be applied to different regions, different vineyards or even different plots of land within the same vineyard.

This is very important because a land's unique terrior is unable to be replicated or reproduced in any other place outside of that area. This is what makes the "place" of where a wine comes from so special.

It is *the* reason why most Old-World wines are named after the place they come from as opposed to the grape varietal. It is a way for winemakers to stand apart from other regions, even sometimes within their larger land holdings, and to show the distinct qualities that no other element in the winemaking process can accentuate.

Terrior might be a French term, but it universally speaks to the taste of the place in the glass transmitted by the grape varietal or varieties. That can happen anywhere from Argentina to Zimbabwe.

Therefore, the environment – the terrior – in which the grapes grow on the vine is, again, *the* most important component in winemaking. Around the world! Period!

It can, however, be a little challenging to describe. It's like one of those, "you'll know it when you try it," type of things.

For example, it becomes as clear as day if you get an opportunity to smell and taste a wine from a certain plot of land in a glass and then immediately smell, feel and experience that same piece of the Earth from your hard.

Each region is equipped with its own geographical gifts that are distinct to that section of land. It's a regional character that shines through. It's the wine's natural identity. It's like the wine having the finger prints of a place all over it.

A chapter focusing on grapes is coming up a bit later. But I'll throw a grape into this scenario to drive the point home a little more.

Let's go with Sauvignon Blanc, for this example. It's a pretty popular white wine grape that can be grown in both Old World and

New World regions. For this instance, we'll focus on a more apples-to-apples comparison: the same grape from generally the same location.

We have two Sauvignon Blanc wines from two different areas of the Loire Valley of France. We have one from a region called Touraine and one from a region called Pouilly-Fumé. The Sauvignon Blanc from Touraine is going to taste very different from a Sauvignon Blanc from Pouilly-Fumé.

Both wines are from the Old-World winemaking region of the Loire in France. Yet the environmental elements of Touraine – the soil, climate, sun exposure – will impart significantly different characteristics into the wine than a wine made from the same grape coming from the nearby Pouilly-Fumé appellation.

The Sauvignon Blanc from Touraine might offer up hints of grapefruit, apricot and green apple while the Pouilly-Fumé Sauvignon Blanc will feature lime, fresh peach and gun-flint notes.

It's the same grape. The grape has been turned into a wine in similar fashion. But the *place* greatly influences the final product. Every parcel, acre or piece of land offers up a different flavor from that one specific place. And that, my friends, is terrior.

Again, it's one of those – "you'll know it when you try it" – situations.

However, let's continue to delve further into how the sense of place can further affect how grapes develop and add to its overall terrior.

For the upcoming section, we'll focus on Burgundy, France. It is considered to have the most dynamic, varied and complex terrior of all the wine regions in the world. Studying the different attributes of this area is like taking a masterclass in terrior.

This section – and book for that matter – is far from a masterclass in terrior, by any means. But to help make the discussion more relatable, I will use examples of Burgundy's terrior when applicable to drive certain points home in the following sections.

: the climate

Climate is king.

It rules everything around the vineyard: the vines, the grapes, the people, the farming, the harvest. When using the word climate, however, we are addressing much more than just weather.

It's definitely a big part of it, but not all of it.

Weather represents the day-to-day conditions of the atmosphere in a region that can fluctuate from minute-to-minute and hour-to-hour. It's the clouds, the wind, the rain, hail, sleet, snow and all that good and not-so-good stuff.

In terms of climate, we want to think more about the long-term weather conditions of a region. When unpacking the topic of climate, the following elements all play a major role in successfully developing our wine grapes: the climate's temperature, the seasons, the landscape, the water and soil.

: the temperature

Temperature is obviously very important to the development of the grapes on the vine. Wine regions need the adequate climate for the grapes to achieve proper maturity by the end of the annual growing cycle.

That is what makes the 30th to 50th parallels in latitude so significant to our wine regions. This portion of the Earth allows for the temperatures to reach an *average* annual temperature of between 50- and 68-degrees Fahrenheit.

These temperatures – ranging from cooler climates to warmer climates – are needed to start the flowering of the vines at the beginning of the cycle, the fruit development during the middle of the cycle and the ripening of the grapes through the harvest.

These numbers, 50- and 68-degrees Fahrenheit, are important numbers to commit to memory. In addition to the average temperatures needed for the growing cycle, these numbers also correlate to the temperatures in which wine should be served and also the temperatures in which bottles of wine should be stored.

You'll learn more about that in CHAPTER SEVEN: **the shopping, serving and storing**. So, remember these temperatures: 50- and 68-degrees Fahrenheit.

Within these the various wine regions of the world, grapes grow well in one of three major climate categories.

The category will either be continental, maritime or Mediterranean. These categories feature distinctive environmental forces that contribute to the ideal development of specific grape varietals. Sometimes these categories can overlap in areas around the world, offering a combination of climate characteristics for certain regions.

CONTINENTAL — This category represents regions that are typically found inland in the wine world. Since these regions tend to be further away from large bodies of water, these regions can experience less rain and result in drier growing conditions. Also, the temperature can go to extremes over the course of a day and a year. During the day, there can be warm temperatures that can take a drastic drop at night. And looking at the overall year, the summer can be pretty hot while the winters can be cold enough for ice and snow. For reference sake, some of these wine regions around the world include sections of Washington State and New York in the United States, Burgundy in France and Piedmont in Italy.

MARITIME — This category represents regions that are typically found much closer to significant bodies of water. The region's closeness to water helps contribute to a steadier, long and even-keeled growing cycle. That leads to warmer summers and cool, but not severely cold winters. Unfortunately, the water can also lead to some potential issues. While the nearby water helps to moderate the region's temperatures, that water can also lead to excessive rain. That can lead to "washed out" flavors in grapes, too much fertility and mildew. These challenges keep winemakers in these regions on their toes. For reference sake, some of these wine regions around the world include sections of Oregon in the United States, Bordeaux in France, Chile, Rias Baixas in Spain, New Zealand and Tasmania in Australia.

MEDITERRANEAN — This category represents regions that are found within the Mediterranean region or exhibit Mediterranean-like characteristics. These areas around the world tend to be warmer than both continental and maritime regions. The growing seasons are long but will exhibit moderate to warm temperatures. Although these regions tend to be near a large body of water, the summers are typically dry, and the winters are considerably rainy. For reference sake, some of these wine regions around the world include classic sections on the Mediterranean basin like Greece and Naples in Italy, and others outside of it like South Africa, Portugal, South Australia, coastal California and the South of France.

Representing general climate themes, these three classifications offer a good overall sense on how the wines will develop. However, to get a closer, more accurate view of things, we have to look at three more categories: macroclimate, mesoclimate and microclimate.

For this lesson, we'll briefly look at Burgundy, France.

Burgundy is a relatively small area. It is located in the northeastern winemaking region of France, just off center and toward the northern part of the country. Some people call it the "Heart of France."

It's stretches from the North to the South, coming in at under 100 miles long. The most northern part of Burgundy is the Chablis wine region. It's a bit of an extension of the Burgundy wine region as it's geographically separated from the main winemaking regions by about 60 to 80 miles. Burgundy has been producing wine since the first century, so for about 2,000 years.

In terms of climate, Burgundy is considered mostly cool-continental with a small amount of maritime, oceanic influences. That means the region experiences extreme weather conditions – cool springs, hot summers, cool falls and cold winters – and a lot of unpredictability. Issues like late frost, hail, drought or heavy rains can severely affect the growing cycle.

Those weather issues are, of course, case-by-case problems depending on where a region falls within the overall Burgundy territory. Those specific issues are also determined from the year-to-year weather conditions just like any other wine region.

One of the most striking things about Burgundy is its terrior. The monks that worked the land the monasteries owned around the 11th century helped create our definition of terrior. That led to the discovery of 100 or so different appellations that showcase the hundreds or even thousands of terrior properties within Burgundy.

This now brings us to the topic at hand, the specific three climate types found in wine regions: macroclimate, mesoclimate and microclimate.

With all the weather challenges that Burgundy faces, there is this almost magical – but definitely scientific – combination of elements that makes Burgundy one of the most distinctive wine regions in the world.

The magic happens when the climate and terrior connect on a more intimate level in the form of macroclimate, mesoclimate and microclimate.

We'll focus on the overall region of Burgundy and drill it down specifically to the Chablis region in the northern part of Burgundy to illustrate these points.

Macroclimate — This is a term that represents the overall climate of an entire wine region. By definition macro means "overall." When looking at the overall region, the annual average climate of Burgundy is between 57- to 61-degrees Fahrenheit. This is referred to as the typical climate of the larger, overall geographic region

Mesoclimate — This offers a lot more focus. It is the term used to describe the climate over a region on a much smaller scale – an area between 30 to 300 feet. This is where Chablis would come into the situation. Since Chablis is a subregion of Burgundy, we would look at what makes up the overall environment that influences the grapes in vineyards there: the altitude, soil types and how near or far these vineyards are from a body of water. For the Grand Cru (great growth) vineyards in Chablis, the vines are planted along the valley of the Serein river. The ideal plots are the ones located on a southwest facing slope that receives the maximum amount of sun exposure. These vineyards are also planted on primarily Kimmeridgian soil, offering up elegance and great structure to the wines. At this level, you start to drill down a specific understanding of a sub region's terrior.

Microclimate — This goes even deeper, looking at the environmental influence on small parcels or plots of land within these sub regions. It could amount to just a few sections of grape vines. One section might be closer to a body of water. One section might get more sun. One section might have a different type of soil. These details will determine the differences in the quality of the wine made from grapes from these highly specific, unique areas. Those conditions can vary from rows of grape vines just feet from each other.

Again, that's a lot to wrap your head around. But that should help explain why there are so many different options of wine on the market. Wine comes from a lot of different regions and sub regions and vineyard plots in the wine world. The goal is to show off the grapes in these unique parcels of land and showcase the very specific environmental forces that shape them.

: cool versus warm climates

Understanding the climate type of the vineyard is one thing and knowing the terrior is another, but what does that ultimately mean for the grapes?

Various grape varietals have individual needs when growing on the vine. Some grapes mature better in a cooler climate while other varietals need warmer climates to fully ripen.

A major strength in successful winemaking is understanding the types of climates a region has in order to understand what grapes are better suited for those areas.

We've already brought up the notion of temperatures being different from the north to the south in the Northern Hemisphere. Then we flipped it and reversed in the Southern Hemisphere with regions from north to south.

To keep things simple, we'll focus on the Northern Hemisphere and those regional examples showcased before: Champagne in the North at the 48th parallel and Mount Edna on the Italian island of Sicily located at the 37th degree.

We'll start with cool climate regions like Champagne, and also Burgundy, since we've been traveling down that path.

Some grapes – red or white – simply grow better in cool climates. They are just naturally better suited for these types of conditions with an annual *average* climate as low as 50 degrees Fahrenheit.

Grapes in these regions don't experience a great deal of continuous heat. The grapes, however, will get all the sun they need to mature.

These higher latitudes vineyards tend to experience longer daylight hours throughout the season.

But the temperatures balance out in a way that allows the grapes to mature on the vines in a steady, consistent manner by the time harvest comes around for picking.

In some cases, like in Champagne, the grapes grow to become barely ripe or even slightly under ripe in these climates. That provides the wines with a high level of acid and a lower amount of sugar. This allows for a certain kind of structure to develop, which we'll jump into more in the next chapter.

The resulting wines from these areas tend to be light, elegant and delicate, but with strong structural components. The wines also tend to come off as a little brighter and crisper due to their growing conditions, typically showing off a great deal of the vineyard's terrior.

Then there are some grapes that love the heat. Not hot temperatures per se, but definitely warmer than those previously mentioned regions. Some grapes just grow better in a warmer climate, regardless if the grape is red or white.

Varietals from the lower latitude regions are associated with the warmer areas and experience early ripening due to the warmer climates, like the Mount Edna region in Sicily. The *average* annual climate in these regions top out at around 68 degrees Fahrenheit.

These warmer temperatures tend to shorten the grape growing season. And if not watched carefully, the grapes can become overripe. That leads to more powerful and robust wines touting richer textures, more fruit notes, a higher level of alcohol and an acid level that is softer.

Let's now bring in a grape varietal that really speaks to a region's climate. You're probably familiar with Pinot Noir. It is a lovely, prestigious grape that makes some of world's most beloved wines.

It's also the primary red grape in Burgundy, where it originated. Pinot Noir will be discussed in more depth in CHAPTER FIVE: **the grapes**.

This grape is also what's called an international varietal, meaning it can be grown in many other parts of the world. However, it can't be planted and grown *successfully* in every wine region.

Pinot Noir is a delicate grape varietal. It is difficult for grape growing experts to farm this grape as it's finicky and very persnickety. Pinot Noir needs a long, cool maturation period in which to grow and develop into its true potential. Therefore, it is better suited for cool climate regions.

In some cases, it can be planted in the cooler temperature areas of a warm-climate region in higher elevations. For now, to keep confusion down, we'll just generally refer to Pinot Noir as a cool-climate grape varietal.

Growing conditions that are not ideal for Pinot Noir – too hot, too wet – can result in a number of issues. Some of those include the wine not offering enough acidity and showcasing way too much alcohol.

That might not seem terrible when you imagine this hitting your palate, until it actually does hit your palate. Then you'll see, first-hand, why the right temperature is so important starting from the vineyard.

: the harvest year

The growing cycle and harvest happen every year in vineyards. Grapes start to bud, develop and ripen on the vine during particular times of the year.

The harvest is the time when grape growers and winemakers decide when to pick their grapes to either sell to winemakers or making wine themselves. Some New World winemaking regions call it the "crush" because that is when they jump into the actual process of making their wine.

Spoiler alert. This section will go back to the topsy-turvy topic of Northern and Southern Hemisphere. Feel free to re-review that part again or just prepare yourself for potentially more confusion.

In the Northern Hemisphere, the grapes grow on the wines from the spring to the fall. That timing is typically the middle of March at the beginning of the season to anywhere from August to early October at harvest time when the grapes are picked. Therefore, in the Northern Hemisphere, if we are in the year 2019, the growing season begins around March 2019 and can go until about October 2019.

Those wines would be considered the 2019 vintage from the 2019 harvest. A wine could be available on the market as early as November 2019, like in the case of Beaujolais Nouveau. But more commonly wines will start appearing on shelves as early as Spring 2020.

Again, things are reversed in the Southern Hemisphere. This can throw many people off.

In the Southern Hemisphere, south of the equator, where the wine regions stretch from the 30th parallel in the north to the 50th parallel in the south, the growing season goes from the fall to the spring.

You see. It's flipped around.

That timing is typically around the middle of September at the beginning of the season to anywhere between February and April for the harvest when the grapes are picked.

In the Southern Hemisphere, if we are in the year 2019, the growing season begins around September 2019 and can go until about April 2020. Those wines would be considered the 2020 vintage from the 2020 harvest. Remember, the harvest is the time when the grapes are picked from the vines and turned into wine. Those wines could be available on the market in the United States as early as the Summer of 2020.

That can be confusing for shoppers because you might be in the year 2020 and see a wine from that same year. It's just because it came a wine region in the Southern Hemisphere.

There is one more thing to note on the growing seasons. The reason why there are such wide ranges in terms of when the harvest starts is because it depends on the climate categories and the season's weather conditions.

Some regions are in cooler climates that will allow the grapes to stay on the vines longer, pushing the harvest back until about October in the Northern Hemisphere. Then some regions are warmer, creating the conditions for grapes to ripen faster which can shorten the growing season and move the harvest up to August or September in the Northern Hemisphere.

Hopefully that cleared up some confusion about Northern and Southern Hemisphere growing seasons and each of their harvests.

Now on to the full growing cycle where we'll look at it through the lens of the Northern Hemisphere to keep it simple and keep us all on the same page.

: the growing cycle

Since our dynamic planet Earth takes 365 days to rotate around the Sun on a tilted axis of 23.5 degrees, we are blessed with our lovely four seasons: spring, summer, fall and winter.

Each season represents an important time for the vines. They are all instrumental in the development of the grapes and the overall health of the vines. However, each season also faces its own unique challenges. Those really depend on the location of the vineyard throughout the wine world. Issues ranging from late frost, hail, drought or heavy rains and being attacked by pests can severely affect the growing season.

With variables like the changing weather conditions and the range of different grape varietals planted, each stage can vary in its length of time leading up to the culmination of the harvest.

We'll take a brief glimpse into those seasons, starting with spring.

the spring

This time of the year is a re-awakening of the vines. They have been resting over the winter after working so hard the previous growing cycle.

With the proper vineyard maintenance and pruning prior to going dormant, the vines are ready to wake up and prepare themselves for growing grapes.

Regardless of what area the vineyards are located in the world, the temperature in the region has to reach a minimum of 50 degrees Fahrenheit in order for the grape growing season to get started.

This is a critical season that kicks off the growing cycle. When that temperature is achieved, the process begins when the vines show signs of "bleeding." It sounds weird, but this is when the ground warms up and pushes water up through the shoots and the water starts to seep out of the "pores" and cuts from the pruning process. Once that happens, very small buds on the vine develop and begin to show and produce shoots. After shoots begin to grow from the buds and the buds eventually break, they become vulnerable to the elements.

The grape growers and winemakers must look after these buds because issues like excessive water can lead to mildew issues or even frost that can put these little buds in jeopardy of being able to produce fruit.

the summer

The temperatures are obviously getting warmer now. The days are longer. The summer months bring around the perfect temperatures for the grapes to start to develop and ripen on the vines.

Now the grape growers are in it to win it and their motto is "slow and steady wins the race." Ask any winemaker or person who grows grapes and they will tell you that the ideal factors needed to grow the best grapes are sunny days and cool nights. They are hoping for the right amount of sunshine in the day and the right temperatures during the night. They need that balance.

The sun, through photosynthesis, helps provide the leaves on the vines with the proper energy to create the sugar in the grapes. This also helps to pass along color and flavor.

With the longer days, the grapes can get up to seven hours of sunlight per day. Daytime temperatures could reach well into the 80s and 90s for ripening. Then at night, the temperatures can drop to around 60 to 70 degrees. The cooling affect helps to preserve the freshness of the grapes and balance out the acidity and sugar levels.

That balance is called the diurnal shift. It's a drastic shift in temperature from day to night to ensure for proper ripening day-to-day throughout the summer season. Those fluctuating temperatures also help each region reach its *average* climate for the year.

By the time May and June come around, the buds have morphed into small, hard grape berries on the vines. Then by July and August the berries grow larger, soften and shift into their intended colors.

The ideal *average* summer temperatures hover around the 70-degree mark Fahrenheit. These temperatures are an important range to stay within as it allows the grapes to stay on the vines longer. This slow ripening, or extending growth on the vines, helps retain acidity, which adds to the intensity, complexity and stability of the wines.

Unfortunately, this time of the year makes the ripening grapes more susceptible to pests and animals. Winemakers have developed tools over the years, both traditional and contemporary tactics, to keep the vines protected from pests, weeds and other issues that might negatively affect the vines and grapes.

the fall

The anticipation is starting to build. All the hard work over the spring and the summer months will all pay off soon – fingers firmly crossed.

Anxiety also sets in because this is a time of great uncertainty. Unexpected weather conditions can swoop in and force grape growers and winemakers to make some fast decisions to protect the harvest. Therefore, they are consistently monitoring the vines, the fruit and the land – checking the weather reports and watching the skies.

As the season progresses towards its end, the winemakers start to make their decisions on when exactly to harvest the grapes. Those decisions are based on the ripeness of the grape, which depend on the desired style or styles of wine they are making.

Grapes are typically picked about 100 days after the vines have flowered. The grape bunches are typically picked in the early morning hours when the temperatures are cooler. The grape growers and winemakers also avoid picking grapes on nights with a full Moon.

the winter

This is the perfect time for a break, well, at least for the vines. Maybe not so much for the grape growers and the winemakers. The vines have worked so hard over the course of the year. They have been prodded. They have been picked. They have been pruned.

They have done their job. And aptly enough, it is time for them to rest and recuperate. After the grapes are picked during the fall, the vines will still pick up fuel from the sun to store through photosynthesis.

The leaves change color and eventually fall off the vines. Once the first frost comes along, the vines enter their dormancy period.

It is this time that winemakers take other steps to help protect the vines during these months to help ensure their health leading up to the next growing cycle.

: the good years and the bad years

"Was this a good year for wine?"

I get that question a lot. And I mean, almost daily.

A "good year" means that all the factors were present that were needed to create the best grapes possible that year. A "bad year" would mean that the conditions were not ideal during the growing season, therefore not producing the quality of grapes to make the best wines by harvest time.

Was 2016 a good year for wine?

Maybe!

What country or region of the world are we inquiring about?

Not sure? Let's stick with France.

The 2016 year might have manifested in the ideal growing conditions in Bordeaux in the southwest region of France, but in Burgundy, located in the northeast of France, maybe not so much.

A "good or bad year" doesn't properly represent a whole region or country. It becomes more of a blanket statement. That conclusion of "good or bad" has to be paired down to the mesoclimate and microclimate of a vineyard in *each individual* wine region. For those determinations, you'd have to research and understand what specifically happened to that individual region during that particular year.

There are lots of variables when it comes to wine, as you are probably seeing more and more as you move through this book. A "good" or "bad" year really depends on the conditions of the growing seasons leading up to the harvest.

: the vineyard environment

By this point, the sense of place and the overall importance of the region should already start to sink in.

Grape vines, much like us humans, are products of their environment. And those environments are made up of so many connecting pieces. Within the larger scope of climate, these contributing elements are water, wind, sun, elevation, slopes and soil.

These represent the natural landscapes within the grape growing areas that greatly contribute to what's going on above the ground and what's happening below the ground.

Again, some of these topics have been brought up before. But I want to shed some additional light on these topics to further demonstrate how they relate to each other.

: the water

It's a no brainer that water is essential to every life form on our Blue planet. In terms of the growing cycle, water has a positive effect on grape development in several ways.

First off, you need a water source in order for grapes to grow. Duh, right!

But it's estimated that a grapevine needs about 28 inches or so of water during the annual growing cycle to help sustain the grapes on the vine. That either comes from rainfall, run off water from high-elevation mountain tops or – where it's allowed – the irrigation of vineyards by grape growers.

The second thing about water, and a very important part of the equation, is that large bodies of it have a moderating effect on the climate. Those large bodies could be in the form of an ocean, sea, lake or river.

Since water changes temperature much slower than air, it maintains temperatures for a longer amount of time. That goes for hot and cold. If the air temperature falls to a cooler temperature, the water is able to release its stored heat to warm things up. Then if warm air rises from the ground, it is tempered by the cool air flowing from the nearby water.

In essence, water has an equalizing effect on the temperature of a region to help serve the needs of the vines and grape growing.

The cool air from the water also helps maintain the needed acidity levels in the grapes during the warmer summer months. Then the air from the water warms the region to protect the vines from freezing roots in the winter and late frosts in the spring.

This is the reason why it's ideal for most vineyards to be near a large body of water. The further vineyards are from the water – like the regions that fall into the continental climate category – the more extreme and challenging the growing condition will be during the season.

But there is also a chance water can have negative effects on the vines. Excessive amounts of water can make the grape vines too fertile. That can create too many grape bunches. That might not sound like a problem to us. But when it comes to winemaking, that leads to major concerns like the possibility of producing bland wines from "watered-down" grapes.

Then, of course, there are the issues of mildew, fungus and other vine-related diseases caused by the exposure to too much water. This is an extremely serious situation to watch out for as that can potentially destroy the crops.

: the wind

As we just learned, the flow of air that sweeps over and through the vineyards can have moderating effect on the vines. It's really fascinating to see how these things naturally relate to each other. But just like water, wind can be a double-edged sword when it comes to vineyards.

Wind definitely plays a role in the overall production of grapes and the wellness of the vines. As water sets on the vines, winds can have a drying effect on the leaves and the grapes. That helps guard against potential fungus-borne diseases. And those winds, particularly at night and in the winter, can protect the vines from extreme cold and frost.

Keep in mind that double-edged sword.

If the winds are too cold, this can slow down the ripening of the grapes. Additionally, extremely high winds can really damage the vines at various stages, from the budding to the ripening process.

This is the case in parts of Southern France when it comes to the infamous Mistral winds. This is a wind phenomenon that affects regions ranging from Provence, the Rhone Valley, northeast sections of Languedoc and the Italian island of Sardinia.

The Mistral is an aggressive and cold northwesterly wind with average strengths of 41-miles-per hour. The winds max out around 115-miles-per hour in the winter and spring. And they typically blow for a few days.

This can have a big impact on the vines and buds during the spring time. It goes without saying that those high, strong winds can blow grapes right off the vines. When the winds max out, they can sometimes even pull up vines from the actual ground.

The positive attributes are that the Mistral winds can blow away much of the cloud cover over vineyards. That allows the necessary sun light to shine on the vines. It also has that moderating effect on climate, bringing cool or warm temperatures up from the Mediterranean. Then in relation to those damp conditions, the strong winds can offset and prevent grape and vine rot and mildew.

While the overall occurrence of winds can have both positive and negative effects on grapevines, they can be mitigated somewhat by location, topography and the use of natural and man-made windbreaks.

: the slope

Aspect. Slope. Orientation. Elevation. These are all words that involve the up or down and the high or low of where a vineyard is placed in any winemaking region.

These factors play an important role in the wine world as they help provide ample access to the sun, cooling conditions and access to just enough water to facilitate grape growth.

It appears Romans farmers and winemakers developed this concept a couple thousand years ago. The winemakers from this era were floating around the concept of terrior before the monks officially landed on its official meaning. The Romans discovered that steep hillsides or mountain ranges are ideal areas to plant wine grapes.

For one thing, vines need a nice dose of sunshine. The amount of sunshine a vineyard gets depends on its aspect to the sun. That's the sloping direction that can face to the north, south, east or west.

The right amount of sunlight depends on the grape varietal and the overall climate of the area. Therefore, the proper slope aspect has to be selected to best suit the needs of the grape varietals in each vineyard.

To paint the picture of aspect through words, we have to look at the wine world again from a Northern and Southern Hemisphere perspective.

I know it seems things get tricky whenever these sections are brought up in the book. Just bear with me. I'll try to make this topic as streamlined as possible.

In the Northern Hemisphere, the north-facing slopes have less direct sunlight, so these areas tend to be cooler. Plantings on this type of aspect would be ideal for warmer climates to help retain the freshness and acidity of the grapes.

While we're still in this hemisphere, south-facing slopes have more direct access to the sun throughout the day. With this scenario, let's also bring in the east and the west. Southeast facing slopes will catch the morning sun. This is ideal to prevent mold, drying out any moisture or dew that might be on the vines from overnight. Then the vines on the southwest facing slopes will get more of the sunlight during the afternoon. This can be ideal for grapes growing in the slightly cooler climates which can help prevent frost.

That is a lot to take in, so I won't complicate the matter further by diving into the Southern Hemisphere's slopes. I will just add that in the Southern Hemisphere, the aspects are flipped.

While the slope is important for the ripening of the grapes, it is also important to help moderate the temperatures in the vineyard. Remember we need two major elements in the wine world for great grapes: sunny days and cool nights.

Going back to the Romans, they learned that cool air runs downhill. When this happens, the cool air settles at the bottom of the valley. Then warmer air comes in at night to add needed heat.

Then there is the matter of water. Just like the cool air flows downward, so does water – particularly excess water.

Previously in this chapter, we went over the negative aspects of too much water in a vineyard. Basically, too much water in the vineyard is a no-win situation for the grape growers and winemakers.

Vineyards planted on slopes provide better drainage. The slope helps prevent the buildup of excess water by encouraging the water to run down the hillside or terrain.

However, in some vineyard areas, like Argentina and Chile in the Southern Hemisphere, farming is assisted from the flow of water from these higher elevation, high-altitude regions.

See how I did that! I had to show the Southern Hemisphere some more love. It got kind of glossed over a bit earlier.

In the Southern Hemisphere, some of the major winemaking regions in Argentina and Chile are located on the foothills and valleys of the Andes Mountains. The Andes have some of the highest mountain peaks in the world. It also happens to be the longest mountain range in the world, stretching about 4,500 miles.

Chile's wine regions fall in the central valleys on the west side of the Andes, while Argentina's regions fall in the varied foothills on the east side of the mountain.

The Central Valley areas of Chile benefit from the rivers that flow down from the melted snow caps of the Andes. That's very helpful as this particular area of the wine world typically has dry growing conditions.

Similarly, in Argentina, more specifically the Mendoza region, the region is also very dry and desert like. Being on this side of the Andes, Mendoza is shielded from rain that would typically come from the Pacific Ocean.

But since the vines are planted in higher elevations, normally from about 2,000 to 5,000 feet above sea level, it allows the vineyards to take advantage of the natural irrigation of the melted snow running down the mountain. That water flow was aided by canal systems built by Incan farmers from the native Andean civilization around the 1400s.

Therefore, in these two examples, you have the run-down water to bring necessary fertility to the land in two different ways. But the sloped vineyards help to keep the water from resting and making the regions too fertile.

: the soil

You have officially landed.

All the other sections of this chapter have officially led us to the more literal sense of place: the soil.

This is a crucial element, without a doubt, when it comes to growing grape vines. The vines literally take root and dig deep into the Earth's soils around the world. We couldn't have one without the other.

But the relationship is sort of rocky. I'm sorry. I couldn't resist. And it's rocky, in more ways than one.

In order to produce quality grapes that turn into good, great or world-class wines, the vines have to struggle. That is the philosophy of the grape growers and the winemakers across the board.

It might sound harsh. It might even sound cruel, but the winemaker is actually doing what is best for the vine.

This concept has been learned through thousands of years of trial and error when farming grapes. Think of it as a healthy dose of "tough love." There's that saying that "what doesn't kill you, makes you stronger." This is very true of the wine grape vines.

In extremely fertile soils, the vine's grapes can grow like wild fire. There can be way too many grape bunches on the vine, which spreads the vine's resources too thin.

If the vines are overcrowded with grapes, there are just not enough natural resources to go around to develop properly. When this happens, the grapes are bland tasting and the wine produced from those grapes will be bland as well.

Soils with lower fertility capabilities make the vines fight for survival. The vines work hard to dig deep into the Earth looking for water and nutrients so that they can survive. That fight makes them stronger and allows them to produce less fruit, but better-quality fruit.

Soils should be deep enough to allow for proper root growth. If the roots are deep into the Earth, that can help protect the longevity of the vine as the roots will have access to water and nutrient reserves well below the surface layers.

But just to clarify things a bit: lower fertility does not mean low- or poor-quality soil. That's not the case at all.

Quality soil is one hundred percent important to successful grape growing. It's great for the health of the vine. It's great for the health of the roots. It's great for the overall wellness of the vineyard.

It's just that the soil can't make the grape growing process too easy for the vines. Great soil and a good fight make for a winning combination. Remember: "What doesn't kill you makes you stronger."

When mentioning specific soil types, there are several different and very important ones relating to grape growing.

Most commonly you'll find soils having large deposits of clay, sand, gravel, chalk, limestone, flint, schist and sometimes volcanic ash. But they can also include terms ranging from alluvial and calcareous to loam and marl, which can consist of a mixture of several soil types.

Within these soil types, minerals are found. All rocks are made out of some type of minerals. Therefore, soils that are abundant in minerals tend to allow more complex aromas and flavors to develop when the grapes are turned into wine.

It helps bring out certain types of "minerality" in the wine, the essence of minerals in the taste and smell of wine. These types of soils are important to the vineyard because they properly nurture the vines all year round – helped by the minerals, nutrients, acids, salt deposits, proteins, nitrogen, carbon as well as the microbiological activity within the soil.

Then the soil needs to have a texture that allows it to breathe. Yes, breathe. It needs sufficient oxygen for roots to breathe and grow.

That happens naturally to some degree but is helped by grape growers and winemakers plowing the land to aerate the soil.

Water is very important, as previously mentioned, so the soils should feature good drainage so the water can go deeper into the Earth so roots can access it throughout the year. And the soils should also be able to retain water, especially in drier areas when rain is scarcer.

The soil can also help with moderating the temperature of the vine's roots underground. It can warm or cool the vines. Some rocky soils are great for retaining the heat reflected on them during the day to keep the vine's roots warm at night. Then those rocks can cool off at night and help cool the vine's roots during the day until they start to warm up under the sun.

That is the official dirt on soil.

Now, let's see how all these elements come together under the umbrella of grape farming.

: the viticulture

The wine world is basically farming. It's agriculture. In the case of grape farming, it's called viticulture: vine agriculture.

But true to form, along with the rest of this chapter, that singular definition doesn't quite do the full meaning of the word justice.

There is one very important element involved: The human element.

Last, but not least in this chapter, we get to bring into the picture the group of individuals who take the time, energy and personal care of nurturing these vineyards.

These are the farmers. The grape growers. The winemakers who own land, rent land and grow their own grapes.

Throughout the chapter, I have referred to these individuals as both grape growers and winemakers. These two groups of people are not always one-in-the-same.

There are some farmers who grow grapes, manage the vineyard and harvest, and then sell off their grapes to winemakers. They are sometimes referred to as a vigneron. That's a grape grower or wine grower.

Then there are those farmers who own or rent land and take care of every aspect of the process, from nurturing the land, growing the grapes and ultimately making wine.

The latter group is commonly referred to as a grower-producer because they handle both the growing of the grapes and the producing of the wine.

I just wanted to make that distinction.

However, whatever their title or role is in the process, there is an important tie that binds them together.

It's love!

Outside of the powerful presence of climate and the intimate influence of terrior on the vineyards, there is a strong connection to the place and the people cultivating the grapes that has been forged since the beginning of civilization.

These professionals have a strong relationship with the land. Regardless of if the farmer's family has lived on the land for generations or if it is their newly adopted land, they develop a dynamic chemistry with the land.

They nurture it, listen to it, feel it, honor it. Over time, these individuals develop a full and complete understanding of the needs of their grapes, the vines and overall land.

But what exactly does viticulture entail?

Viticulture requires closely monitoring and assisting the vines throughout the year. During the winter, they prune and protect the vines while they rest during the dormant period. This is the time to get them ready for the spring flowering period, fertilizing the vineyards and watching out for issues like frost and other problems. Once the grapes develop, it's all about maintaining them during the ripening period. This time is crucial as it's important to guard against diseases and pest. That eventually leads them to the harvest when they decide the appropriate time to pick the grapes.

If the viticulturist, the grape farmer, is also the winemaker, then they continue on the path to produce their wine, having a full, first-hand understanding of the growing cycle's details. If the viticulturist is growing the grapes to sell to winemakers, then they must be in constant contact with the winemakers throughout the growing cycle and the harvest.

Ultimately, it is up to the viticulturists to decide how they will care for the land. They determine how they will deal with the Earth's changing weather patterns in the form of climate change and global warming. They figure out how they will manage pests. And if the land shows signs of illness of any kind, it will also be up to the viticulturist to decide if they will continue with traditional practices or if they will be inspired to mix things up and change perspective.

: the organic viticulture perspective

Long ago and many, many moons before chemical-based herbicides, pesticides and fertilizers were dreamt up in a scientist's mind, winemakers found natural, organic ways of managing issues related to fertilization and pest control.

Organic farming is typically regulated by the government and uses a variety of natural preventative treatments that are believed to have little or no negative impact on the environment. That means steering clear of harsh chemicals and additives related to either the grape growing, the winemaking or both. The goal is to help minimize the erosion of the soil, promote biodiversity in the vineyard and protect the overall vineyard environment.

Some winemakers opt to become certified by the government or a certification organization to assure customers that their products are indeed organic. That can be a long and expensive process, particularly for smaller producers. Therefore, many farms will make their wine in an organic manner, as a normal business practice, and opt out of becoming certified for a label designation.

I know some of this information spills over into actual winemaking as opposed to grape farming, but there is such close overlap that I wanted to address it all together to help shed light on this category and clear up some consumer misconceptions.

There can also be some confusion, unfortunately, when it comes to the labeling of these wines. Some bottles will state "made with organic grapes." This is a different requirement than organic wines. It is a less strict way of making the wines under the organic umbrella; only the grapes have to be certified organic and not the actual winemaking process.

More confusion sets in when some "organic wine" labels will state "no sulfites." That, unfortunately, perpetuates consumer misperceptions of wine in general. *All* wines have sulfites. That's a natural part of the winemaking process. Sulfites will be discussed more in the next chapter. However, there are no *added* sulfites in organic wine. Organic wines just have the naturally occurring sulfites that are found in all wines. But not adding additional sulfites is a way to help keep the wine in its more natural state with the least additives possible.

Since the production requirements are different with wines listed to be "made with organic grapes," sulfites can be added or not to these wines. That decision is up to the wine producer.

: the biodynamic viticulture perspective

Every time I think of biodynamic farming and winemaking, I think of the song "Sister Moon" from the 1987 Sting album "Nothing Like the Sun."

The opening lyric asks the "Sister Moon" to be the singer's guide.

The beginning of that song kind of perfectly sums up biodynamic agriculture because grape vine planting and harvesting are based on the moon's cycle, aka, the lunar calendar.

This style of farming was developed in the 1920s by Rudolf Steiner, an Austrian philosopher, scientist and spiritualist. In 1924, Steiner started to give a series of lecture to farmers in Europe who were concerned about the future of agriculture and the well-being of their land.

Steiner suggested that the farmers develop an alternative relationship with nature. He advocated for highly organic, natural ways to manage the vineyard. Think of it as a homeopathic way of treating the vines.

He promoted taking into consideration the overall ecosystem of the region and vineyard, using astrological influences for farming. This takes into consideration the push and pull effects on the Earth from the sun, stars and moon that give life to the land, vines and grapes.

Without going into too much specific detail because this is the "less is more" approach, it is essentially a holistic and ecological – and some believe moral – approach to farming, gardening, food, wine and overall nutrition.

The wines produced in a biodynamic style are produced very much like organic wines. There are no added sulfites. And the focus is on allowing the wine to become what it is destined to be by nature and not by mankind.

The concept has had a resurgence within the last twenty years or so, with even a selection of wine producers from Champagne, the South of France and even one of the most expensive, exclusive and well-respected Bordeaux producers, Domaine de la Romanée-Conti (DRC), using the biodynamic practice of farming.

That reinvigorated interest has also spread to other wine regions around the world and in other various facets of agriculture.

With the concern for what chemicals have done to the land and farmer's concerns about the future of agriculture, it could – at some point – become the norm and not the alternative.

Only time will tell.

: the sustainable viticulture perspective

Viticulture is without a doubt a labor of love. As mentioned before, there is an affection for and a loyalty to the land. Farmers are intrinsically linked to it. It's like a personal extension of them. With that amount of care and attention, there comes a heaping dose of concern and worry these days.

Farmers are fearful that the worst-case-scenario could quite well become a tragic reality. That is that their land and their industry won't survive for future generations.

This is where sustainable farming comes into the picture. Sustainability looks at the long-term ecological, economical and socially responsible practices of farming.

These farmers want to remedy mistakes made from the past and look into regenerative agriculture methods. They are thinking about the greater impact farming is having on the world and finding ways for the land to sustain itself for years to come.

Sustainability could involve organic or biodynamic farming practices along with energy conservation, packaging, air quality, water quality, transportation emissions, the surrounding wildlife, the people working at the farm and/or winery and the communities it serves.

At the heart of it, sustainable agriculture is about being a much more responsible and better corporate citizen with eyes set firmly on the future.

CHAPTER THREE: **the elements**

Wine is one of the most delicious puzzles on this planet Earth.

What is a puzzle if not a tool that challenges our brains, thought processes, perspectives and also triggers our excitement?

Its various elements call on our curious nature to think and ponder and wonder how these pieces intrinsically link together.

It also encourages deductive reasoning and adaptable thinking that lets us focus on very nuanced elements in order to deepen our comprehension of the matter at hand.

Puzzling, right!?! That is wine!

There are lots of tiny little pieces that have to be assembled in order for wine to exist as we know it.

But before we dump out all the pieces of this puzzle, we first have to understand what wine is exactly.

Wine is defined as a fermented beverage produced from the juice of any fruit, most commonly grapes. The Vitis vinifera species of grapes in this case.

When we talk about wine in the everyday sense of the word, we are referring to what is known as "table wine." These are still, non-sparkling wines that are normally associated with wines you'll have on their own or with meals.

To start with, we'll be introduced to the major players in the process and we will go over how wine is made. We'll move on to the elements and structures of wine. Then we will see how these pieces relate or don't relate to the five major styles of wines featured in this chapter: white, orange, red, rosé and sparkling wines.

Ultimately this chapter will attempt to put those puzzle pieces into perspective to allow you to understand its full, enigmatic picture – faults, flaws and all.

And it all starts with the transmutation of grapes to wine.

: the winemakers

At the helm of this metamorphosis from grapes to wine is the winemaker. The technical name for these professionals is an oenologist, the practice of oenology or the science of winemaking.

Regardless of what they call themselves – winemakers, oenologists, wine growers, vintners, vigneron, négociant – they are my personal modern-day heroes and "sheroes."

I mean, of course, besides the living legends that inspire the heck out of me every day like Tina Turner, Barack Obama, Rita Moreno, Cicely Tyson and the Notorious RBG (Ruth Bader Ginsberg) to name a few.

But yes, to me, these men and women are heroic figures from all over the world.

The reason why I feel this way is because they are such an integral piece to the winemaking puzzle – past, present and future.

These ladies and gentlemen have dedicated their lives to personally taking on the, sometimes thankless and heartbreaking, job of keeping this ancient tradition of wine production and consumption alive. And not only just keeping it alive, but pushing the industry forward, finding new ways to overcome challenges and serve as a sort of midwife to help deliver wine to us.

In addition to the winemakers being brilliant scientists, I like to think of them as amazing artists-in-residence. There is a sensitive yet majestic dance with nature they must carefully choreograph to get the best out of the grapes they turned into wine.

They must also be proficient at orchestrating all the different facets and players involved in each growing cycle to properly work together. Winemakers must know how to balance out elements in the wine to let its true essence sing out, loud and clear. And they must fashion a wine that will be lovely, dynamic and a respectable representation of the place in which the wine comes.

Their responsibilities can be as different from each other as their opinions on how they personally perceive their role in the winemaking process.

Some like to take a very hands-on approach to winemaking. This group of winemakers, both from the Old World and The New World, strongly believe that the process requires a large amount of human intervention. They believe that winemaking requires extreme precision and technique. So, it is their job to get intimately involved in every facet of the winemaking process while continuously incorporating new learnings and new technologies.

This approach is more academic and scientific in style, from the vineyard to the winery. This group may lean closer to the notion that "a wine made in the winery."

Then there is another school of thought.

Other winemakers take a more hands-off role in the process. These winemakers, both from the Old World and New World, believe in giving the grapes enough room to develop on their own with less human interference. They honor the practices and procedures of their Old-World predecessors, keeping their eyes on past successes and their hands tightly gripped around tradition.

This approach is a more naturalistic, minimal intervention approach, from the vineyard to the winery. This group leans more closely to the notion that a "wine is made in the vineyard."

Regardless of their title, their specific role and personal wine philosophy, love fundamentally fuels their professional passion for winemaking.

: the fermentation

From solid form to liquid form, these vinifera grapes take on a second life when they are transformed into wine.

Grapes, and this species in particular, are ideal for winemaking. That is because everything is contained in and on the grapes that is needed for the fermentation process – converting the grapes into wine.

Those essentials are yeast and sugar. The yeast is found on the outside of the grapes and the sugars (mainly fructose and glucose) are found inside the grapes.

The alcoholic fermentation process is when the yeast eats the sugar and converts that sugar into alcohol. That is how still, non-sparkling wines are made. It's the primary alcoholic fermentation.

But let me back up a bit. I'll take it from the top and go back to yeast to allow the story to unfold properly.

Yeast is a single-celled microorganism that feeds on sugar and simple carbohydrates. We've pretty familiar with the concept of yeast when it comes to bread and maybe even beer.

Yeast is abundant in the atmosphere. It floats around us all the time. It lands on the skins in the vineyard where the grapes grow.

You might have noticed while shopping for grapes – regardless of the color – that they have a powder-like substance on the skins. If you haven't noticed it before, you will the next time you visit a grocery store, farmers market or vineyard. That yeast on the skins is called the "bloom" or the "blush."

This now brings us back to the alcoholic fermentation stated before.

Fermentation happens when yeast interact with the sugar in the grapes. When the grapes are pressed and crushed, the yeast cells from the skins are now physically introduced to the sugar in the juice. The yeast will then eat the sugar, creating heat through this metabolic activity while also converting the sugar into alcohol.

There is a byproduct that is also produced during the fermentation process, carbon dioxide (CO2). However, the carbon dioxide blows off the wine and into the atmosphere during the primary fermentation stage.

This particular fermentation mentioned uses the native yeast to spark the fermentation. There are times when winemakers opt to add yeast to the process. More about that will come back up very soon.

In a passive form, without physically pressing the grapes, that fermentation process can also happen right in your own refrigerator. If the grape skin has a tear in it from damage or because it starts to spoil, the yeast will find its way inside and begin to ferment that grape.

I'm sure we've all tasted a "hot grape" that started to develop alcohol in our fridge at some point. If not, I can just see people putting this book down to poke holes in their clusters of grapes like Mad Scientists to see how the magic works. Or maybe even logging on to Amazon to invest in an at-home winemaking kit.

That is the primary fermentation in which all wines go through. However, some wines go through a secondary fermentation to achieve a certain style of wine.

: the secondary fermentation

As discussed in the primary fermentation process, we lose a by-product of that fermentation to the environment. That product is carbon dioxide. While the yeast eats the sugar, they are not only producing alcohol they are also producing the carbon dioxide and heat during the fermentation process. While a moderate- to high-level of carbon dioxide would be considered a fault found non-sparkling wines, the CO2 has a very special place in sparkling wines.

Sparkling wines feature captured carbon dioxide that provides them with bubbles. The most popular sparkling wine is Champagne from Champagne, France. But sparkling wines can be made from all over the world. We'll jump into more detail later in this chapter.

In order for the winemaker to get the bubbles into the wine, the wine has to go through a secondary fermentation. This fermentation can happen in individual bottles or in a large Stainless-Steel tank. The bottle method is referred to as the Method Champenoise/Champagne Method or Method Traditional/Método tradicional. The Stainless-Steel tank method is called the Charmat method or simply the tank method.

To spark the secondary fermentation with the Champagne Method, the winemaker pours the still wine into individual bottles. Wine, sugar and yeast are then added to the bottles containing the still wine. This mixture of wine, sugar and yeast is called the liqueur de tirage in French. The bottles are then sealed with a crown cap, similar to the caps used to seal beer.

Now that there is sugar and yeast together in a capped bottle, the yeast will feed on the sugar in the bottle and convert that to alcohol, creating heat and carbon dioxide. The carbon dioxide is then trapped in the bottle in the form of bubbles, producing a sparkling wine. The bottle will go through a series of processes before it reaches its final stage – which can take several months or years to complete – and is released to the market.

The Stainless-Steel tank or Charmat method involves putting a large amount of still, fermented wine into a large pressurized tank or vat. Additional wine, yeast and sugar are added to the tank to spark the secondary fermentation. There are a few processes before the sparkling wine is bottled and is sold to consumers. But this process is pretty quick and streamlined. It can take a few months to complete.

That is the secondary fermentation that creates the popular sparkling category.

The next secondary fermentation is used by select winemakers to affect the taste and texture of wine. That is known as malolactic fermentation. This fermentation process will come a little later in the acid / acidity section of this chapter to make more sense.

: the yeast

Linking back to the historical perspective of CHAPTER ONE, wine has been made for thousands of years before winemakers actually understood how grapes became wine.

It was either considered a gift from the Gods by some or an agricultural mystery by others. What seemed like a mystical and magical mystery for thousands and thousands of years turned out to be pure science.

It wasn't until the mid 1800s when the science behind fermentation was discovered. After the global "Industrial Revolution" which helped lead to the invention of the microscope, French scientist, Louis Pasteur, uncovered the alcoholic fermentation process back in 1850.

The fermentation process mentioned in the previous section was sparked by wild, native "ambient yeast" found in the vineyard that land on the grape skins. These types of yeast help provide wines with more "terrior" or a sense of place in the wine. However, winemakers can opt to use "cultured yeast" for a more controlled fermentation. This particular yeast species is **Saccharomyces cerevisiae** or **S. cerevisiae**. It can also be called budding yeast, brewer's yeast or bakers' years. It's important to call this out by the full technical name because it is similar to the yeast used to ferment beer, bread, chocolate and spirits.

It is an individual decision for the winemaker to use this type of yeast. Various types of yeasts can also add flavor to the wine and some are better suited for different grape varietals than others.

: the fermentation vessels

Evolving from a happy accident to purposeful production, winemakers have opted to use various vessels in which to ferment wine.

This has definitely evolved leaps and bounds from the ancient earthenware that was used during the "Hunters and Gathers" period.

There are three major vessels used today: oak barrels, stainless-steel tanks and clay pots. Winemakers can use one vessel solely or opt to use a combination of these vessels for their winemaking purposes.

Oak barrels – also known as Barriques – have been ideal vessels since the Roman Empire for storing and transportation purposes. But overtime, the use of fermenting and storing wine in oak barrels became the norm due to its unique abilities to allow wine to develop, soften, absorb flavor and age well over time.

Wine barrels can feature *neutral* oak barrels for fermentation and storage purposes. These are barrels that have not been treated by fire on the inside. These types of barrels don't give off any major flavoring to the wine, except maybe damp wood.

For flavoring of wines, toasted oak barrels – new oak or reused oak – can impart toasty, smoky and / or flavorful notes to wine. The wood is charred or toasted on the inside of the barrel sort of like a crème brûlée. When the wood caramelizes, it brings out some natural flavorings like its sugars and a chemical called vanillin, the same compound found in vanilla beans. These flavor and aromatic notes can give some drinkers the impression of cakes or muffins or pastries.

The major styles of oak barrels winemakers use are French, American and Eastern European. Because of its porous nature, the barrels allow for the gradual influx of oxygen. The flow of air can both soften structures in the wines and also make the wine more concentrated through evaporation.

With a typical size barrel being about 60 gallons, some winemakers tend to lose about five gallons partly as it evaporates into the air – called "The Angel's Share" – and also by being absorbed into the wood. After all this time, the wine world hasn't given up the thought that wine is a gift from the Heavens. Some things are still sacred in the world.

The selection of oak used depends on the winemaking country. French oak is used throughout the world because of its elegance. Winemakers love the texture it offers and the subtle flavors of coffee beans, cloves and baking spices it imparts into the wine. American oak can offer up more aggressive flavors like vanilla, dill and coconut. These tend to be favored by winemakers around the world that produce bolder wines. Eastern European oak from Croatia, Hungary and Slovenia tends to be produced in larger formats like Botti and Foudres, sometimes more than 1,000 liters in size, made for the purpose of the extended aging of wine.

The ultimate flavors and aromatics of the finished wine are dependent on the age of the barrel and its toast levels.

Most red grapes benefit from oak influence, yet it can have a dynamic effect on neutral-flavored white grapes and richer white grape varietals.

In terms of those white wines that have oak interaction, many consumers believe that a "butter" taste is derived from oak barrels. But that is not a byproduct of oak. That is a part of a secondary fermentation in wine. Again, there will be more details about that in the upcoming "acid" section of this chapter.

Some winemakers choose to cut costs by soaking oak chips in wine resting in stainless-steel tanks to obtain that "oaky" flavor. That proves to be a much less expensive alternative to purchasing oak barrels.

In recent years, there has been a big consumer backlash against the use of oak and oak barrels in some wines. The presence of oak in some wines can be overpowering and unattractive to some wine drinkers. The opposition against oak by some wine lovers has prompted some New World winemakers to produce "unoaked," "unwooded" and "naked wines." Those wines are free of any oak aroma, flavor and influence of any kind.

Stainless-steel tanks might be the new kids on the block in terms of fermenting vessels, but its introduction revolutionized the wine industry globally from the New World to the Old World.

Springing forth from another happy accident in history, stainless steel was developed when Harry Brearley of the United Kingdom happened to add chromium to molten iron in 1913 to ultimately create what we know now as stainless steel.

Around that same time, refrigeration started to slowly become the norm in businesses and homes around the world. Fast forward to the 1940s, the use of refrigeration in winemaking started to be incorporated in winery rooms. By 1964, the development of temperature-controlled, stainless-steel tanks allowed winemakers to have more control over the fermentation process of wines – especially white wines which seemed to damage more easily during fermentation.

Stainless-steel tanks – also known as stainless-steel vats – are large, neutral vessels used to ferment wines. Being a neutral vessel, these tanks don't offer any flavoring or coloring to the wine. The tanks allow the essence and integrity of the grape to stay intact and shine through in the wine.

Since the tanks proved to be easier to clean, the vats also helped insure fresher, cleaner wine by preventing bacteria from spoiling wine and creating off-putting smells. The end result is a more aromatic, less stringent wine that is ultimately better in quality.

Clay pots have been used in winemaking from the Neolithic period by ancient civilizations in China, Egypt, Greece, North Africa, Armenia and the Republic of Georgia. With the popularity of the "Natural Wine" movement, new styles of clay pots are making a comeback among a handful of Old and New World winemakers. What was old is new again.

Essentially the pots of are Terracotta ("Baked Earth") made from clay. While the term "amphora" is used commonly to describe the pots, other names like tinajas and dolia – to name a few – represent more specific styles and shapes of these clay pots.

These pots were used primarily to store and transport wine and other goods in ancient times. Amphora served as an ideal fermentation tool because it's a neutral vessel that doesn't impart any flavor into the wine, just like Stainless Steel tanks. However, the wine can sometimes seem a bit more mineral driven due to the porous nature of the clay. The pots can allow for a good amount of oxygen exchange, have natural cooling and clarification properties, and are great for aging wine. Some of the newer clay pots are sometimes coated with epoxy. In ancient Greece, the pots were coated with pine resin to slow down the oxidation of wine.

Many modern-day winemakers stretching from Georgia, Slovenia, Italy and Oregon are experimenting with new styles of clay pot fermentation for many of its red and orange wine offerings.

An off-shoot of clay pots are the new temperature-controlled **concrete eggs tanks** used by many winemakers in California and other parts of the world. This style is more similar in makeup to the clay pots, but also features some of the same benefits oak and stainless-steel fermenters offer. The concrete wall keeps the wines cool and can also control the fermentation. The eggs also allow small amounts of oxygen to interact with the wine like oak barrels, but don't offer any flavoring to the wine like stainless-steel tanks and clay pots.

And, that my friends, is fermentation. It's a nice chunk of the wine puzzle. But there is much more. Let's jump into other key components and structures of wine.

: the structures of wine

These are the true puzzle pieces that make up the elements of wine. At the core, all wines will have the overall structures found in wine – alcohol, acid, minerals and tannins. But the structures, and related elements, can vary depending a bit on the style of wine.

We've got a lot to cover, so we'll just dive right in.

: the alcohol

We will start off with alcohol. I can just hear the cheers and snaps and claps as this is a major reason why people drink wine – for the effect alcohol has on them.

All silliness aside, alcohol is truly an important structure in wine. It is literally what separates wine from grape juice.

We just learned about how grapes are fermented and converted into wine. The reason grapes are ideal for wine is because they are high in sugar content from two primarily sources: fructose and glucose. These sugars allow wines to ferment into a desired amount of alcohol.

The average alcohol for wines is between 12 and 14 percent. That is referred to as the alcohol by volume (ABV). Remember that this an average. You will find some wines with an alcohol content as low as 5 percent and some wines as high as 17 percent.

Also, don't forget that alcohol and sugar are linked. The sugar is turned into alcohol in the wine. When the yeast eats the sugar and converts it to alcohol, all the sugar can be consumed to achieve a higher percentage of alcohol or some sugar can be left unfermented in the wine in the form of residual sugar.

It is also important to note that alcohol is a natural preservative that helps give wine its shelf life.

When we look at a wine bottle with a 5, 8 or 10 percent alcohol content, we are looking at a wine that is going to be on the sweeter side. These wines have a decent amount of sugar that was not eaten by the yeast. On the opposite spectrum, bottles with an ABV of 12 to 17 percent have very little or no residual sugar left in the wine. That makes those "dry wines."

: dry versus sweet wine

A dry wine – red, white, rosé, orange or sparkling – means the absence of sugar. Little or no residual sugar can be found in the wine. The opposite of a dry wine is a sweet wine when referring to "table wines."

That is not to be confused with dessert wines or fortified wines that can sometimes reach 19 to 21 percent in alcohol. This book is not focusing on those wines. That is a different category of wine that encompasses styles like Sauternes, Sherry, Port, Tokaji and Ice wine.

There's already so much of information to present in this book, so I've opted not to focus on this category of the wine world. Maybe it'll make its way into the next book. I'm working very hard to stick with the "less is more" theme.

Circling back to dry wines and sweet wines, those are the bookends for table wines. From dry heading to the sweet side, you have wines that are "off dry" with a little residual sugar and "semi sweet/demi sec" in the middle before arriving at sweet. A dry wine can have less than 10 grams per liter of residual sugar, while a sweet wine can weigh in close to 30 grams per liter of residual sugar. This is all attained through the fermentation process.

Looking back at CHAPTER TWO: **the place**, we learned that grapes can grow to be under ripe or very ripe. That ripeness level affects the sugar content. Less sugar in the grapes will result in lower alcohol in the wine if the yeasts finishes it all. A riper grape, bursting with sugar, will turn into a wine with a higher level of alcohol and, in some cases, some residual sugar if the yeast can't finish it all off.

Sparkling wines have a slightly different set of dry to sweet rules, but we'll get into that later in this chapter.

: light, medium and full-bodied wine

The body of the wine – the texture, the viscosity – is an element that people really respond to but sometimes have a difficult time putting those tactile feelings into words.

That is unless, it's a rich, full-bodied wine. A lot of people gravitate toward the full-bodied wines, for tons of reasons. But there is a wide spectrum of textures associated with wine.

Essentially, the body of the wine is the weight and the texture of the wine on your palate. In the wine world, we tend to liken the body of wine to the body (weight and texture) of milk. It makes sense to do that since most people have had exposure to milk in their lives at some point.

We use the comparisons of skim milk, 1- and 2-percent milk, and whole milk. I haven't gotten to taste test soy milk, almond milk, rice milk or hemp milk to gauge all their textures just yet. Maybe I'll do that in an e-book or as a YouTube segment. But, for now, we'll go with the traditional milk categories.

Starting off with light-bodied wines, we liken the weight and texture of these wines to the weight and texture of skim milk. It's very light and lean on your palate. Those options would be associated with wines with low- to low-to-moderate alcohol levels.

We'll jump over to full-bodied wines next. Full-bodied wines have a generous amount of weight and texture to them. We liken that body of wine to the body of whole milk. There's a lot more richness and viscosity there. Those wines would represent wines that have a generous volume of alcohol.

PUBLIC SERVICE ANNOUNCEMENT: Be cautious of wines with a 15 percent or higher ABV. I know some people drink for effect, but those higher levels can make the alcohol seems "hot" or overly aggressive if any other structure of the wine is unbalanced. If a wine is unbalanced, that means one of the structures of wine stands out more – most likely in an unpleasant way – than the others.

Now back to our regularly scheduled program.

Then we'll go back to medium-bodied wines. Medium-bodied wines are more closely associated with 1- or 2-percent milk. Now, I don't expect everyone to know the varying degrees of the textures of milk from skim to 1- and 2-percent milk to whole milk. I'm still trying to get better acquainted with the milk alternative textures. But a 1- or 2-percent milk is not as thin as skim milk and not as rich as whole milk, so it falls right between those two as a point of reference. The alcohol will more moderate in scope for these wines.

: legs

This brings me to one of the most frequently asked questions about wine: the legs.

People have been commonly told to look for a wine with "Great Legs." They were told to analyze the "legs" on a wine by swirling it in the glass because it meant that the wines were better quality.

Not to be a Debbie Downer or Party Pooper, but legs – thick or thin – do not equate to quality in any way in a wine. I don't think they equate to quality in a person either, but that's another type of book all together.

The only thing legs, also called tears, really tell us about wine is its thickness. The more defined the legs and the slower those steams run down the glass, the thicker the wine with a potential for more alcohol and residual sugar. Lower ABV percentages, even with some residual sugar, will have less of a coating effect on the glass and produce thinner or nondescript legs or tears.

If the way you determine a wine's quality is by its alcohol content – and I've met several people who do – then in that sense it would be deemed a high-quality wine based on that characteristic.

: the aromas

Wine can offer up a myriad of scents within the glass that allows us to connect with what we are experiencing through our noses. It's quite amazing!

When it comes to being introduced to the smell of various wine grape varietals, you are awaking your smell memory to a smell, or collections of smells, in which you've already had some contact. The smell of a wine arouses emotion and pleasure that can possibly connect you with past memories that have been filed away in your smell memory.

Maybe the smell of rose petals in a Nebbiolo wine brings back memories of your grandmother who loved wearing rose perfume. Does that New-World Chardonnay remind you of all that ambrosia salad you ate as a child, with its creamy hints of coconut, marshmallow, citrus and pineapple?

The aromas in wine are specifically connected to individual grapes. But the available options of smells expand when you factor in the way the wine was made and where the grapes were grown. Although these are grapes, when fermented they take on different chemical compounds that are shared with other things, like the rose and coconut and citrus from the previous two examples.

Dr. Ann C. Noble, a sensory chemist and retired enology professor from the University of California at Davis (UC Davis) took a scientific approach that best describes the smells found in wine. That led her to develop the Aroma Wheel. She related the specific chemical formulas associated with scents in wines and associated them with the common nouns we use on a regular basis.

For example, the chemical formula for citrus acid is C6H8O7. Most people wouldn't understand what C6H807 means. I had no idea. I don't speak that language. But I do know what citrus – lemon, lime and grapefruit – mean and smells like. Therefore, it's a matter of using your sense of smell while adding your language to describe what you are smelling in the glass.

The Aroma Wheel breaks down wine aromas into 12 basic categories that are further divided into different aromas that fall into those main categories. There are primary notes you can get directly from the grape through the fermentation process like orange, cherry, blackberry, vegetables or lemongrass. Then you get secondary characteristics through the winemaking process like smoke, toast, vanilla, coconut and butter.

Aromas can also be shaped by the age of the wine. When a wine is younger, it offers up more fresh fruit notes from sour to sweet. Then as wine ages, those fresh fruit notes tend to turn into more concentrated notes of aged fruit, like a fruit that is almost past its prime or a fruit that has been dried like a raison, apricot or fig.

When my students suggest they are not good at smelling, I tell them to just concentrate on their smell memory: what they've been exposed to over their life. I encourage them to not focus on what they think they should smell, but what they actually smell based on what's in the glass. This practice of smelling the wine will be fleshed out further in CHAPTER FIVE: **the sensual approach to wine tasting.**

: the acid / acidity

Acidity is an important structure in wine that is not often as appreciated as alcohol. Many wine drinkers are reminded of a sharp feeling and / or tangy flavors when tasting a wine with a high acid content. That is true in some regards. However, this section will delve into expanding the overall notion of acid and acidity in wine.

Acidity is a key factor in maintaining the quality of the wine by helping to preserve its freshness over time. Acid, like alcohol, is a natural preservative.

There are three major acids found in wine: citric acid like what is found in citrus fruits, malic acid like what is found in green apples and tartaric acid found in grapes.

The acidification of wine starts with the grapes on the vine. As we discussed previously in the last chapter, there are some cooler climate wine regions and some warmer climate wine regions.

In the cooler climate regions, let's say like Champagne, France, for example, the grapes that grow there will oftentimes grow to be under ripe or just barely ripe. That is based on the general cool climate of northern France. Given that the grapes are under ripe or barely ripe, the grapes develop and maintain a high level of acid and not a lot of natural sugar.

In very specific terms, outside of the normal conversation about acid in wines, in goes back to the pH level in the grapes.

Normally, I don't get too deep into the pH discussion in the wine classes. Nonetheless for this book, I think it can help the reader more thoroughly understand the scope of acid in wines. So, we're going to briefly scratch the surface on pH levels.

The initials pH stands for "power of hydrogen." By definition, pH is the measure of the molar concentration of hydrogen ions in a substance. That's basically the level of acid or alkalinity – a chemical measurement of a water's ability to neutralize acids – of a solution. The scale ranges are from 0 to 14.

Unfortunately, there are some concepts related to wine that are absolutely counter-intuitive or contrary to our common sense thinking. The pH level scale is also counter-intuitive. Items with ranges below 7 are acidic solutions. Items with ranges of 8 or higher are considered alkaline.

For example, lemon juice, vinegar and wine all come in under 4 on the scale: 2 for lemon juice, 2.2 for vinegar and 2.9 to 4 for wine. Then water comes in at 7, which is neutral, while the chemical lye comes in at 13 on the scale.

As the pH scale indicates, all wines have acid, from a high level to a moderate level. We really don't want a wine to have a low-acid content because the wine will come across as lifeless and flat on the palate.

When a wine touches your palate, it has a fresh, mouth cleansing quality. Since all wine has acid, that makes food and wine such a great combination. The wine works to either compliment or contrast the dish, while the acid does the heavy lifting of cleansing your palate and getting you prepared for your next bite of food and next sip of wine.

Traditionally sparkling wines, by far, have the highest acid levels. These wines are ideal to either serve as an aperitif or throughout the dinner. Before a meal, the acid cleanses your palate of any previous items you've tasted like gum or coffee. Then the acid makes your mouth water or "salivate," which sparks your digestive system and make you hungry.

There are some red and white wine grapes that produce moderate- to high-acid wines in the form of lighter wines, while some of the richer white and red options will feature and moderate to moderate-minus acid levels.

This is good to know when analyzing the wine by the color, the taste and the texture of the wine.

When looking at red wines, the more acidic wines with lower pH numbers tend to look more like bright cherry red to medium red in color, while those with higher pH numbers are more purple, blue and sometimes blueish grey.

On the palate, higher acid wines tend to be sharp, tart and tangy. Moderate-acid wines come off as a bit softer on the palate, more mellow. And moderate-minus acid wines feel smooth and creamy on the palate.

It is not uncommon for winemakers – especially in warm climate regions – to acidify the wine or add acid into the wine. That process helps add balance to the wine when grapes become over ripe on the vine.

Then there are some winemakers who want to soften the acid in the wine. That finally leads us back to a secondary fermentation called malolactic fermentation, in which some winemakers use to achieve a distinctive style and texture of wine.

: the malolactic fermentation

This particular secondary fermentation option doesn't have anything with bubbles in the bottle. It is all about achieving a certain style of wine that comes off as very rich, creamy and oftentimes buttery.

This section was better suited after the acid section opposed to the fermentation section because it directly deals with converting the sharp malic acid, most commonly found in green apple, into a softer lactic acid found in milk, cream and butter.

The primary role of malolactic fermentation is to deacidify wine or reduce the acid content. The process usually happens directly after the first fermentation but can sometimes take place in conjunction with the primary fermentation.

Malolactic fermentation is a pretty common practice used on select red or white grapes to enhance the flavor and body of the wine, particularly if the grapes were grown in a cool climate region. It is also used with some white wines to impart a creamier texture and the essence of butter – in the form of diacetyl – on the nose and on the palate.

Overall, this secondary fermentation is used by winemakers to make the wine seem more approachable, drinkable and more pleasant to drinkers who prefer a lusher style of wine.

: the minerality

The topic of minerality is – believe it or not – a hot topic of debate for many of us in the wine industry. That basically stems from the disagreements on what "minerality" actually is in wine, where it comes from and if it truly exists.

Minerality in wine is associated with the smell and taste of elements connected with the Earth's ground. That consists of rocks that are wet and / or have been crushed, metals, salinity, dirt and soil. Then there are those who dig a bit deeper into the makeup of the soil, which can include traces of calcium, nitrogen, potassium, magnesium and sulfur.

Regardless of whether or not these minerals from rocks and metals are actually imparted into the grapes, there is no denying that certain soil types like chalk, clay and limestone play a crucial role in the health, wellness and productivity of the grape vines.

I personally don't doubt that minerality exists in wine. My point-of-view, in terms of minerality, is that since these grapes grow on vines in soils from around the world, wines do feature some traces of the minerals. I've smelled and tasted so many wines that give off the essence of limestone and flint and volcanic ash.

The argument could be just an issue of semantics. If the existing scientific evidence is limited on the subject or does not back up its existence in wine, maybe what we describe as minerality is essentially just part of the wine's overall traits as an agricultural product.

While the wine world is thousands of years old, it can continue to be a bit of a mystery to us. There is still a lot left to figure out. I'm sure some well-established conclusions about wine will be challenged by technological advances and new learnings in the science fields.

Time will tell if we get our answer to the minerality question. If or when we get that answer, will it be satisfactory to us? Or will it even matter to those who believe in the presence, existence and contributions of minerality when it comes to the taste and smell of wine?

: the tannins

If there is a question about the existence of minerality in the wine world, there is no questioning the existence of tannins.

You have all probably had this experience before with red wine. You've taken a sip and instantly you feel your mouth start to dry out. It's very real. It is tactile. You can feel it when it hits your palate.

While tannins are a very tangible structure in wine, it can still prove challenging for wine drinkers to describe properly.

First off, what exactly are tannins? Tannins are chemical compounds, polyphenols, that have a natural astringency to them in liquid. Tannins are found in the skin, stems and seeds of grapes. When the tannins meet your palate via the wine, your mouth will dry out and feel dehydrated to varying degrees.

The juice of red wines has a longer interaction with the skins and therefore develop a prominent tannin structure. White wines don't have tannins because the skins, stems and seeds are removed in the winemaking process. Orange wine and rosé wine will have a small amount of tannins, depending on how long the juice sits with the skins.

Tannins can also be found in oak, tree bark and tea leaves, just as another point of reference.

In the wine world, tannins are grouped into three major strength categories: soft (low), moderate (medium) and firm (high).

A wine with soft tannins will have a mild drying out quality on your palate. Moderate will have a medium drying effect. Firm tannic wines will have a severe drying out quality, similar to feeling like you have cotton or gauze in your mouth while at the dentist.

Like acid, another important structure in wine, the tannin structure also helps with food and wine pairing. You've probably heard about this concept of pairing red wine with red meat. It's a pretty commonly used philosophy in the culinary world. It is true that red wine can go especially well with red meat. But it's not as simple as putting the two together and waiting for the magic to happen. There are layers to this pairing, so let's start peeling them back.

The presence of fat can minimize the drying out sensation of tannins in wine. Therefore, if you have a wine with high, firm tannins, that would be better paired with a red meat that is rich in fat and protein. Something along the lines of a Porterhouse, T-Bone or Rib-Eye steak.

Here's how it works. Your steak of choice is cooked exactly how you like. You take a bite. Once you chew on the steak, the fat and protein in the steak will begin to coat your palate. Now your palate has a nice layer of fat and some residual protein. Once you take a sip of your highly tannic wine, the fat will minimize the drying out effect of the wine.

This goes for high-tannic wines and big hearty steaks. If you have a wine with soft, lower tannins, like a Pinot Noir, you would want to opt for a leaner cut of beef like sirloin tip or fillet mignon as not to overpower the wine too much.

The other great thing about tannins are that the polyphenols are a natural preservative. With this extra layer of preservatives – in addition to alcohol, acidity and sulfites – red wines will have a longer shelf life than white, rosé and orange wines. Some age-worthy reds can last several decades when stored properly.

There will be more details about food and wine pairing guidelines as well as storing and aging wine in CHAPTER 7: **the shopping, serving and storing**.

: the sulfites

Sulfites are sulfur-based compounds that are a natural part of the fermentation process. All yeasts will produce detectable amounts of sulfites when it converts the sugar from grapes into alcohol. While sulfites occur naturally in wine, the compounds are also added to wine in some cases as preservatives. To help slow down the oxidation of white wines that lead to the browning and spoiling of the wine, sulfites are typically added at higher levels to white wines. Red wines have the tannin structure which is a natural preservative that slows down that oxidation and spoilage process, but these wines can also have added sulfites. Sulfites also prevent the growth of yeast or bacteria.

Unfortunately, to the disappointment of many who feel sulfites are bad for them, there is no such thing as a "sulfite-free" wine. Consumers who are concerned about sulfites can opt for wines that have "no added sulfites," most commonly found in natural, organic, sustainable and biodynamically produced wines. I'm not a doctor and I definitely do not play one on TV, so I encourage wine drinkers to consult their personal doctor, nutritionist, dietician and / or allergist to determine if they have sulfite allergies, digestive sensitivities or respiratory issues that could be trigged by this element in wine.

: the tartrates and sediment

Sediment and tartrates are sometimes an unwelcome surprise in the glass for many wine lovers. All wines containing tannins are capable of having sediment in the bottle or "throw sediment" as we call it in the wine world.

Sediment is found in the bottle of mostly red wines, but you might get small doses in orange and rosé wines if they are kept around long enough. Since these wines have tannins, those tannins over time will start to form tiny particles within the bottle as the wine ages. Those particles will cling together and then precipitate to the bottom of the bottle.

Again, sediment is normal. It is also harmless. But it can be distracting, unattractive and uncomfortable when it shows up in a glass or on your tongue. Chewing on those small particles can also leave a bitter taste in one's mouth. See CHAPTER SEVEN for details on how to remove sediment before the wine is poured in a glass.

Typically for white wines, some will have sediment-like particles in the bottle called **tartrates**. You won't see that very often. There are, of course, exceptions for this, like white wines like Gruner Veltliner and older Rieslings, for example. These particles are colorless and caused by the tartaric acid in the wine. These are affectionally called "white diamonds" in the wine world. These are normal and harmless, so nothing to fear. These particles can be easily filtered out of the bottle as well.

: the wine faults

Wine has this mysterious, almost mythical strength that makes it a very powerful juice. Not many beverages can age, develop and mature over time like wine. Some wines can last for several decades and still be extremely vibrant and delicious.

It boggles the mind in one aspect. But it makes sense, especially if you subscribe to the notion that wine is a gift from the Gods. But even Dionysus had his weaknesses. When it comes to the major weakness of wine, it is its sensitivity.

That high-level of sensitivity can make even the most robust wine fall completely apart into an unappealing, undrinkable substance. Wine is extremely sensitive to temperature, light, vibrations and odors. With having these sensitivities as living, breathing entities, wine will become susceptible to problems from time-to-time from production, bottling and storing mistakes.

These issues can really taint the experience for wine lovers. If you purchase a bottle that has one of the following flaws, contact the wine store in which it was purchased. Most wine stores will not offer refunds due to wine laws. But many will replace the bottle for you if the bottle has been damaged by a wine fault.

Here are some common faults that can develop overtime that affect wine's overall character.

Cork taint — The first major issue is the risk of TCA taint or "cork taint." I'm really trying to just give you the essentials in terms of information, but I got to give you the full name on this one. TCA is short for 2,4,6-trichloroanisole. Less is more, right! This is a fault that occurs in wine when the TCA chemical compound leaches unfavorable smells and flavors from the cork into the wine. That contamination, although not dangerous to consume, gives the wine "off" aromas and flavors associated with soggy cardboard, mildewed towels or a moldy basement. It's estimated that between 3 and 8 percent of wine with cork stoppers suffer from cork taint or are "corked" as we say in the business.

Bacteria — In this case, there could be a number of reasons a bottle can be spoiled by bacteria. After some microbial exchanges, there could be some left-over bacteria that has a chemical reaction with other compounds in the wine. That could lead to an unintended second fermentation or leave the wines with foul smells and tastes. Unsanitary conditions in a winery can lead to bacterial growth that will negatively alter the smell and taste of the wine.

Brettanomyces (Brett) — A little "Brett" can be okay for wine drinkers in some circles, but too much can be considered a major fault. Brett is spoiled yeast. In small doses, it can offer up smoke and leather on the nose. But in larger doses it can be funkier expressions like a Band-Aids, barnyard and must. It can affect all wines but is more of a problem in red wines.

Sulfur Dioxide — When added, sulfur dioxide acts as an antioxidant and an antiseptic. In high doses, unappetizing smells from the sulfur compounds can include the aromas of rotten eggs, burnt matches or canned asparagus.

High alcohol — While there are some wine lovers who will drink solely by the high alcohol content stated on the bottle, too much of a good thing can be quite bad. If you choose wine to consume one evening, you don't want to feel like you just had a tequila shot. If the wine has a high alcohol content, and that content is not in balance with the other structures, the wine will appear "hot" on your palate. That represents a strong burning sensation in your mouth caused by too much alcohol in the wine.

Low acid — If too much alcohol in a bottle of wine makes the wine feel "hot," too low of an acidity level will make the wine feel "flabby" or "lifeless." All wine should have a decent amount of acid. Low acid happens if the grapes are not picked at the right time in the vineyard, especially in warmer climate regions. Low acid means the wine is not balanced, or not in complete harmony, with the other structures in wine.

Volatile Acidity — If the acid is way too sharp, you have a wine that is on the verge of spoiling or becoming oxidized. Some experts say that wine's overall destiny is to turn into vinegar. What they are referring to is the acetic acid in the wine, often referred to as volatile acidity (VA) or vinegar taint. This acid, commonly found in vinegar, sharpens when the pH level of the wine lowers to about 2.2 on the scale. This type of acid would be the polar opposite of a low-acid wine. This fault can have an extremely sharp bite on the palate.

Over Oxidation — Oxygen is a friend and an enemy of wine. A nice dose of oxygen or proper aeration of the wine can allow the wine to stretch and relax and mellow out. But too much oxygen will lead to spoilage. An overly oxidized wine can have a sharp acid content with dull or muted flavors. Sometimes that happens in the bottle before it is opened. Many times, if will happen after the bottle is open for a few days, as the wine begins to spoil with prolonged contact with oxygen.

: the styles of wine

The picture of this wine puzzle is almost complete. We had to build a framework in which to fill in these last missing pieces. Most of the hard work is done now and we are just to the point where we can start realizing the full picture.

The last few pieces of this puzzle revolve around the various styles of wine. I'm referring the major categories of wine: white, rosé, red, orange wine and sparkling wines.

These categories are seemingly straightforward. But with anything and everything in the wine world, we've learned that things are not always so cut and dry, black and white or red and white.

I'll divide this section into still wines and sparkling wines.

Let's first get into still wines. Still wines are essentially the everyday regular table wines we drink on a regular basis: meals, binge watching, happy hours, socializing. They are called still wines because there is no carbon dioxide or carbonation in the wine.

This is our major wine category.

In terms of colors, Mother Nature conjured up a brilliant palate of shades. In a glass, you can view everything from the palest straw and pink in color to copper tones and the deepest reddish-brown hues.

Before we take a look at specific grapes in the next chapter, we'll explore these larger wine categories one-by-one.

: the white wines

Despite being clear, bright and offering up a range of colors from greenish-yellow to deep golden, white wine can still prove quite perplexing for many wine lovers.

Made by grapes that have green, yellow or even pinkish-grey skins, white wine is very distinctive. It's the only category in wine that has very little or no skin extended contact with the juice. As a result, the pulp on the inside of the grape takes center stage and is allowed to shine through in the wine.

To make a white wine at the end of harvest, the grapes are picked and then pressed and crushed. When the grapes are pressed and crushed, the juice from the inside of the grape is extracted. The juice is extracted into one vessel after being pressed, while the skin, stems and seeds are left behind as the wine production progresses.

The yeast particles attached to the outside of the grape will eat the sugar in the juice and converts it to alcohol. And voila, you have a white wine.

White wine has been made for centuries. However, research suggests that this style of wine became very popular after World War II thanks to the introduction of refrigeration in winemaking and Stainless-steel tanks. These tanks allowed winemakers to have more control than before over the fermentation process of wines. White wines can damage easily during fermentation. It was easy to pick up unattractive flavors and aromas before this technology was developed.

By focusing on the flavor and aroma compounds located in the pulp of the grape – forgoing a tannic structure – winemakers were able to create wines full of youth and vibrancy that can be enjoyed while the wine is young.

However, with the wide range of white grape options and the textures that can be achieved, white wine can range in color, texture and alcohol. Therefore, white wines can be fermented in Stainless-steel tanks to be young and fresh as well as being treated by oak barrels developing rich, complex and age-worthy options.

Then, of course, there is the curve ball. Just when you think you have white wine all figured out, there is something else to process.

While we know winemakers can make a white wine from the vast selection of white wine grapes, they can make a white wine from red wine grapes as well.

What? Yes! I know. Crazy, right!?!

The pigment of wine comes from the skins. We'll discuss that more in the next few categories. Therefore, removing the skins from red grapes will put the focus on the pulp on the inside of the grape which is the same color of *most* grapes.

That style of wine is called Blanc de Noir. It's a French phrase that means white from black. This category, however, is pretty rare when it comes to still, non-sparkling wines. It does exist for table wines but is more common for sparkling wines.

White wines can fall into all three bodies styles: light, medium and full. That's dependent on the grape, winemaking style and aging.

Wine lovers like to enjoy these wines chilled and/or cold, oftentimes without food. These white wines, while they can be enjoyed on their own, will also make great companions with a variety of food items.

: the red wines

This is a category in wine in which people tend to have found a nice comfort zone. While finding an ideal white wine to satisfy their palate might be a more laborious process, many wine lovers can find more immediate satisfaction in a number of red wine options.

Red wines are made from grapes with a red, blue or dark-purple color skins. The red grapes are pressed and crushed, then the juice sits with the skins, stems and seeds. The mixture of the skins, juice and seeds is referred to as the "must." Many times, however, the stems are removed from this process. All these elements are left in direct contact each other for an extended amount of time.

The color of the skins bleeds into the juice. The overall time depends on the grape and how much pigment, flavor and tannic structure the winemaker wants to extract from the skins. The juice can be stored with the skins for a few days or several months.

The color of the red skin grapes is related to a couple different factors. If a red-skin grape is grown in a cooler climate, the grapes don't get a constant source of heat and sun. Therefore, the skin on the grapes would be thin and with more pulp inside the grape. When the grapes are pressed and crushed, the must is left to soak – macerate – with the skins. Since the skins are thin and there is higher level of pulp, the red wine will have a lighter color and more transparent appearance.

When red skin grapes get a substantial amount of heat and sun from warmer regions, the skins on the grapes become much thicker and there is less pulp on the inside. These wines tend to showcase a much deeper, darker color and are sometime opaque appearance.

Then there is the matter of age. Young red wines are typically vivid with colors of bright reds, blues and purples. As the wines age and mature over time – in the cask or in the bottle – the wines start to lose their bright colors and take on more reddish-brown, brick and brown colors. That's just a sign of oxidation over a period of time. It's like what happens to lemon or lime wedge over time. It starts to brown with extended contact with oxygen.

Red wines can fall into all three bodies styles: light, medium and full. That's dependent on the grape, winemaking style and aging.

While red wine can be enjoyed on its own, the experience will heighten when the wine is matched with ideal food options to bring out certain flavor components. Food tends to help temper some of the structural elements of red wine like higher alcohol and tannins.

: the orange wines

Orange wines are a throwback to winemaking in times gone by that have made a resurgence in today's wine world. Utterly unique in smell and flavor, some wine lovers wonder, however, if this type of winemaking should have been left long forgotten.

This category represents an ancient way of wine making, dating back some 8,000 years in the early days of winemaking. And it was still a very common wine style up until the early 1960s in countries like Italy and Slovenia.

Orange wines are made from white wine grapes. To make an orange wine, white skin grapes are made like a red wine. That means the skins, seeds and sometimes stems are left to sit with the juice of the grapes for an extended amount of time. The length of time with the skins and juice could be one day or a few months.

As we've just learned, wine gets its color from the skins of grapes that leach into the juice. The colors that white skin grapes give off with extended juice to skin contact are ranges of orange, copper, amber, pink and hibiscus. The skin contact also gives orange wines a slight tannic structure. That structure brings in a certain level of astringency to the wine. It also brings an extra layer of natural preservatives, giving orange wines a much longer shelf life than white wines.

In the sommelier world, from wine nerds to cork dorks, these wines are all the rage at the moment. Visually the color is striking and, oftentimes, the wine appears to be a little cloudy or murky. On the nose, these wines can smell reminiscent of hard apple cider, whiskey, an oxidized fortified wine like sherry or something that resembles a hot toddy mixture. The palate reinforces those oxidized and alcohol notes, while having a muted, dried fruit flavor – potentially with some floral notes and a generous amount of minerality.

For the general public, some consumers just don't see the fascination. That is very understandable as orange wines are a big shift from the clean, bright and fresh white wines they are more accustomed to drinking.

Having a resurgence, orange wines – in part by the natural wine movement – are being produced in places like Croatia, Italy, Slovenia, California and New York. Many producers are also opting to go back to more ancient ways of fermenting the wine in large clay pots or temperature controlled concrete egg tanks.

Orange wines fall mostly into medium- and full-bodied styles based on the winemaking style and aging expectations.

Love it, hate it or indifferent to it, orange wines are truly quirky. They also happen to work really well when paired with food, especially some of the more challenging food options or dinners with a vast array of dishes, like buffet or family-style dinning.

: the rosé wines

There's probably no other category as explosively popular at this time than rosé wine! If the orange wine category is more niche, the rosé category is definitely the mainstream. It just gets more and more popular each summer.

Like red wine, rosé wine is made from grapes with a red, purple or dark-purple color skins. There are two distinct methods to achieve a still, non-sparkling rosé:

Limited maceration — After being pressed and crushed, the skins, seeds and juice only sit together for a short amount of time to achieve a light bleed of color. That time frame is most commonly between 2 and 48 hours in stainless steel tanks or a neutral oak vessel. Two hours of skin contact creates a light-colored wine. Around 48 hours, you have a wine that has deeper, darker tones. Some winemakers will opt for a little longer skin to juice contact for longer than two days. It is important to note that darker rosé wines are not automatically sweeter. These wines will offer more body but will still fall on the dry wine spectrum – unless the winemaker opts to make the wine sweet or semi-sweet.

Saignée method — This is a method that produces a rosé wine as a byproduct of red wine production. Saignée means "bleeding" in French. While producing a red wine, winemakers will bleed off a portion of the juice. This portion will be more rosé in color as the wine hasn't spent much time soaking with the skins. This method happens fairly early on in the winemaking process. The goal for this method is to concentrate the texture, flavors and structure of the red wines being produced. Therefore, it produces two wines, the red version and the rosé version.

The color of rosé wine can range from pale pink, onion skin, rose water, rose gold, salmon and medium ruby. These wines will have a very subtle tannin structure with its limited skin to juice contact.

An important distinction must be made when it comes to this particular color category, however. In the United States, there is a style of “pink wine” that is referred to as “blush” wine. That is different from traditional rose. Blush wine is made in the U.S., most notably California. These styles are made to be semi-sweet or sweet. That would be like a White Zinfandel or White Merlot, for example. Rosé wine comes from around the world, the majority of which are fermented to be dry. But there are some lovely sweet rosé options as well from France and Italy.

Rose wines can be made out of any red wine grape and therefore fall into all three bodies styles: light, medium and full. That’s dependent on the grape, winemaking style and aging expectations.

Consumers enjoy drinking rosé wine on its own, particularly in the warmer months. However, rosé can be enjoyed all year round and is quite lovely with a variety of foods.

: the sparkling wines

This is important. Please be sure to make a special note of this. It just might be on the final exam!

Champagne is a sparkling wine, but not all sparkling wines are Champagne. That very important distinction has to be made.

Sparkling wine has bubbles caused by trapped carbon dioxide in the wine. Some people call it bubbly wine. Some people call it fizzy wine. But the general category across the board – from around the world – is called sparkling wine.

Champagne is a region in France in which some of the best sparkling wine hails from in the world. Only sparkling wine from the Champagne region of France can be called Champagne. Champagne represents a place in the world and also a style of making sparkling wine. All other sparkling wine – outside of Champagne, France – has to be called sparkling wine.

This goes back to the importance of why the “place” of a wine is so important. The region of Champagne represents a distinctive terrior. It is very different from most wine regions in the world due to the climate, soil types and small number of grape varietals allowed to grow there. Champagne, therefore, offers up a distinct “sense of place” that only Champagne, France, can provide.

Confusion sets in among consumers within this category when people use the branded and regional name, Champagne, to refer to all sparkling wines from around the world.

To help curb the confusion a bit, there are some specific names for sparkling wines from different regions. Knowing these names tend to help wean people off of using the blanket “Champagne” name for all these wines.

Sparkling wine made in France, but outside of Champagne, France, are primarily called **Crémant**. In the South of France, other names can apply like either **Blanquette** or **Mousseux**.

In Spain, these sparkling wines are called **Cava**.

Italy has a lot of sparkling wine options throughout the country. The word “**spumante**” is the Italian word for “sparkling wine.” In the Veneto region of Italy, these wines are mostly called **Prosecco**. In Lombardy you have **Franciacorta**. Sweet sparkling red, rosé and white wines are produced in the Piedmont like **Brachetto d’Acqui** for red or **Moscato D’Asti Spumante** for rosé and white options. Then **Lambrusco**, a red, white or rosé sparkling wine, is produced in Emilia-Romano.

In Germany, the sparkling wine is called **Sekt**.

In South Africa, the category is called **Cap Classique**.

And, again, the words “sparkling wine” serve as a general catch-all for other wines with bubbles.

Even with all those different distinctions, a large population of consumers will still refer to all sparkling wines as Champagne.

Part of the confusion is because Champagne became very popular and well respected globally through marketing efforts. The word Champagne carries so much cache that it eclipses the names of most of the other styles. Many customers don't even know there is a distinction between the names and styles of these sparkling wines.

The other part of that confusion was ushered in when United States sparkling wine producers started to use the word "Champagne" on their bottles. This actually sparks major confusion among U.S. consumers to this very day.

It's a fairly long story. The short of it is that European Union (E.U.) has long tried to protect the name Champagne from being used internationally by companies outside of that region. There was a treaty signed by the E.U. and American winemakers, The Treaty of Versailles, after World War I, stating that only sparkling wine from Champagne could be labeled as "Champagne."

The treaty was signed by the U.S., but never ratified by the Senate. That left a loophole in the agreement which was revisited during another wine-trade agreement in the early 2000s. This time the two countries agreed to forbid the use of the word "Champagne" on any new wines produced in the U.S. However, the agreement "grandfathered" select companies who had already used it on their packaging. These producers were allowed to continue placing Champagne on the labels with a modifier, like California Champagne or New York Champagne.

That was just a little more historical insight for you.

Now let's focus on what makes sparkling wine so special: the bubbles. During the fermentation section of this chapter, secondary fermentation for sparkling wine was introduced. It's a method used by winemakers to trap the carbon dioxide which creates the bubbles in the juice.

In order to produce any sparkling wine, a still, non-sparkling wine has to be made first. That wine could be white, orange, rosé or red. We have gone through that traditional alcoholic fermentation a few times. The yeasts on the skin eats the sugar in the juice. That process creates heat and eventually carbon dioxide and alcohol.

During the first fermentation process, we lose the carbon dioxide. In order to capture the carbon dioxide, the wine must undergo a secondary fermentation initiated by the winemaker. There are a few different ways to accomplish that.

THE METHODE CHAMPENOISE

The leading secondary fermentation style is associated with Champagne winemakers. It has been crafted over hundreds of years of trial and error. It is called the **méthode champenoise** (Champagne method) or the **méthode tranditionalle** (traditional method). This is simply a secondary fermentation that occurs within each individual bottle of sparkling wine – regardless of what region or country the wine is being produced.

The way that secondary fermentation happens, however, is not so simple. The base wine that was created is poured into individual sparkling wine bottles. The base wine can be made from one particular grape or a blend of grapes. It can also be the same wine from a blend of different years or vintages. The winemaker can make 600 cases a year or they can make 600,000 cases a year, but the base wine has to be put into each individual bottle for this process to occur.

The winemaker then adds a mixture of wine, yeast and sugar to the base wine and caps the bottle with a crown cap, similar to those used on beer bottles. While in the bottle, the yeast will eat the sugar (because that's what yeast do) and the secondary fermentation begins. The process creates heat and adds alcohol and carbon dioxide to the wine, which is now trapped in each individual bottle.

The bottles are then placed on a riddling rack to rest at a 35- to 45-degree angle. The winemaker then has a person, team or machine softly twist the wine bottles a little every day for three months or up to one full year. That allows the yeast cells, which have flavor components, to integrate well with the wine. It also shakes off sediment from the sides of the bottles. This process is called riddling or remuage in French. The goal is to make sure the yeast cells rest in the neck of the bottle.

To remove the dead yeast cells from the bottle, the winemaker dips the neck of the bottle into a brine solution at temperatures below the freezing mark (-16 degrees Fahrenheit) or they use liquid nitrogen. Once the liquid in the neck quickly freezes, the winemaker turns the bottle right-side-up and pops off the crown cap. The pressure from the bubbles in the bottle pushes out the frozen sediment full of dead yeast cells.

While we're on the topic of pressure. The pressure of the bottle from the carbon dioxide can be quite a lot in sparkling wines. It has been measured as the equivalent of two car tires, as high as 6 or 7 atmospheres of pressure or it can be as low as 1 to 3 atmospheres of pressure. Sparkling wines that are fermented in the bottle tend to have a high amount of pressure, making the bubbles in the wine more prominent.

To finish off this process, the wine is then topped off – in most cases – with a wine and sugar mixture to achieve the winemaker's ideal dryness or sweetness level. This process is called the dosage.

That brings us to the dry to sweet discussion. We've learned that dryness in wines means the lack of sugar. A wine with little to no sugar in the juice is considered dry. Still wines with some residual sugar can range from off-dry to semi-sweet. Then wines with a lot of residual sugar would be considered sweet.

Hold that thought for a while. The same concept applies for sparkling wines, except there are specific terms associate with sparkling wines. And sparkling wines have different residual sugar levels amounts than still, non-sparkling wines to differentiate dry from sweet.

Here are the general levels from the dry-to-sweet categories in sparkling wines:

BRUT NATURE / BRUT ZERO — This sparkling style is the very driest, with very little sugar found in the bottle. This can have up to 3 grams of residual sugar per liter. The juice really shows off the terrior or place in a clean, fresh way. There are a small number of producers that make this style, so it is not very common to find these readily available on the market.

EXTRA BRUT — This represents a very dry sparkling wine. The residual sugar only tops off at about 6 grams of residual sugar per litter. There is a very low amount of sugar added during the dosage process. This style is a little more readily available on the market.

BRUT — This is the most common dryness level of sparkling wines found on the market. This category represents wines with up to 12 grams of residual sugar per liter. Many of the Brut styles fall within the 7 to 12 range. This is still considered a dry sparkling wine. If 12 grams seems like a lot of sugar, compare that to about 90 to 125 grams of residual sugar for a cola or lemon-lime soda.

EXTRA DRY — This is when the discussion of dryness levels in sparkling wine gets confusing. Extra Dry represents between 13-17 grams of residual sugar per liter. As with some other subjects in the wine world, this concept is counter-intuitive. When "dry" in still wines means little to no sugar, the word "dry" in sparkling wine represents more sugar. Historically, sparkling wines were a lot sweeter in the 1700 and 1800s than they are today. You will find some Prosecco, Cava and U.S. sparkling wine options on the market with this dryness level on the label. These will have a higher perception of sweetness on the palate than the previous three categories.

SEC — Not found a lot on the market, the Sec category – which is also called "dry" – represents 18 to 32 grams of residual sugar. Again, this harkens back to a time when these wines were produced to be sweeter because of taste preferences and preservation needs.

DEMI-SEC — This is more commonly found on the market when it comes to the "sweeter" styles of sparkling wine. The demi-sec category has between 33 and 50 grams of residual sugar in the wine. These styles will give you a nice dose of sweetness. These sparkling wines are often enjoyed with desert or after dinner as desert.

DOUX — The sweetest of them all, the doux category comes in at a whopping 50-plus grams of sugar. These are not common on the market and represent a segment of drinkers that historically enjoyed much sweeter sparkling wine. However, you might find this style of sweetness with the sparkling Moscato options. But you never know – the "what's old is new again" concept might bring this one back to the mainstream.

Is your head spinning yet? And not from a sugar rush! Or was it a sugar rush of sorts?

After the desired dryness or sweetness level is achieved by the dosage, the bottles are then sealed off with a mushroom cap, a wire cage and left to rest on their sides in the wine cellar until the bottles mature and are ready to hit the market.

The other aspect of sparkling wine you want to consider is the vintage. The majority of sparkling wines from around the world are in the form of non-vintage sparkling wines. That is when winemakers blend multiple vintages of wines together – reserve wine with a fresh vintage – to create their signature house style of sparkling.

Then in select cases, sparkling wine will have a vintage if the winemaker deems that a specific year was ideal for growing grapes. When that happens, the winemakers will only use grapes from one particular year to make this sparkling wine. That, therefore, makes it a vintage sparkling wine.

This practice of producing “vintage” wines is associated with sparkling wines from around the world. However, vintage sparkling wines are rarer and regarded to be better quality than their non-vintage counterparts. As a result, they will also be more expensive. A perfect example of vintage Champagne is Dom Pérignon, the prestige cuvee from Moët & Chandon. The Dom Pérignon brand only produces vintage Champagnes and consequently is rarer, more expensive and highly regarded. Since 1921 until now, the brand has only made about 42 vintage Champagnes from very specific years.

Wines that undergo this style of in-the-bottle fermentation include Champagne, Crémant, Cava, Franciacorta and sparkling wines from many other regions around the world like Germany, South Africa and the United States.

CHARMAT METHOD

Another popular way of triggering the secondary fermentation in sparkling wine is called the **Charmat method** or the tank method.

This process was started by an Italian inventor, Federico Martinotti, who developed and patented his original idea in 1895. Then the concept was finalized by French inventor Eugène Charmat in the early 1900s.

The method might carry the name of the French inventor, but the process is tied to sparkling wine made mostly in Italy. Like the traditional method, a base wine must first be created. The wine can be made from one particular grape or a blend of regional grapes. The base wine is put into a large Stainless-steel tank that is pressurized. The winemaker then adds a mixture of wine, yeast and sugar to the base wine, which kicks off the secondary fermentation.

The tank is sealed, and the carbon dioxide is trapped in the wine. The yeast is filtered out. The wine is bottled. These wines rest very briefly, then they are released into the market.

Sparkling wines that are fermented in the tank tend to have high to moderate amounts of pressure in the bottle, from 4 to 6 atmospheres of pressure. These styles can be quite bubbly. Then there are also styles that fall between 1 and 3 atmospheres of pressure. These wines are considered slightly effervescent or fizzy; **frizzante** in Italian or **pétillant** in French.

Wines made in this style typically come from Italy in the form of Prosecco and Lambrusco. But this method is also used for affordable sparkling wines from the United States, Germany and other places around the world.

METHOD ANCESTRAL

Once you have lived long enough, you'll see that the "what's old is new again" concept becomes a constant reoccurrence in the life. The wine world is no exception.

The only difference is that the wine world has enjoyed a long-life span, so what is old in the wine world is often ancient.

We saw that in the resurgence of amphora clay pots. We saw that in the resurgence of orange wines. We saw that with a resurgence of biodynamic farming. And now we see that with a special interest in sparkling wines that are made in the **methode ancestrale** (method ancestral/ancestor method) style.

Long before Dom Pérignon was tasked with the job to solve the problem of the refermentation of bottles found in cellars in Champagne, France, back in the late 1600s, monks in the South of France were working with sparkling wine. The South of France is reported to have been where monks discovered that still wine can turn into sparkling wine back in 1531. That is 107 years before Dom Pérignon was even born.

This is said to have been the world's first sparkling wine. Or at least the first documented source. And in keeping with that heritage and legacy, select sparkling winemakers have continued to create wines in this style. Although, the method is said to have evolved a bit over time.

The steps that separate the ancestor method from the Champagne method is that there are no disgorgement and dosage processes. Therefore, the wine is left with its sediment, yeast cells, in the bottle and there is no additional wine and sugar mixture added to set a desired dryness or sweetness level.

Basically, half way through the fermentation process in a larger vat, the wine is added to each bottle and sealed off. The wine continues to ferment in the bottle, capturing the carbon dioxide. Since these don't undergo the disgorgement process, these wines are technically unfiltered and have traces of yeast sediment.

The pressure in the bottle varies on these types of wines. Some can get up to 4 atmospheres of pressure, while many tend to be fizzy, frizzante or pétillant in style.

Wines that undergo this style of winemaking can come from areas in France like Jura, Loire Valley and Limoux. But there are many winemakers from Italy, Slovenia and parts of the United States, from California to New York, producing these styles. This style tends to coincide with the slow wine or natural wine movement. Under that movement, many of these styles are referred to as Pét-Nats (Pétillant Natural) wines.

: the colors of sparkling wine

Sparkling wine is so festive. It's not only because of the fun bubbles, but because it can come in all of the major wine colors: white, orange, rosé and red.

The most common color is white wine. Since the art of making orange, rosé and red wine exists, we can also find those with bubbles. Here are some of the major styles you might encounter along your sparkling wine journey.

WHITE SPARKLING WINE — This style is achieved by making the wine out of white grapes or red grapes or a combination of the two. The term for sparkling wine that comes from only white wine grapes is called **Blanc de Blanc**. That means white from white in French. Of course, that's pretty common place in the wine world. Then there are white sparkling wines made using red-skin grapes. The winemaker removes the skins immediate after pressing and just uses the inside of the grape to make the wine. That process creates a **Blanc de Noir** sparkling wine, a French term that means a white from black. These white sparkling wines can give off a variety of flavors from fresh citrus, jasmine, stone fruit, apple, grilled pineapple and toasty notes. The can also range from dry to sweet.

ORANGE SPARKLING WINE — These styles of "orange" sparkling wines come in the form of Pét-Nats. These wines are made from white wine grapes that are left to sit with their skins and sediment, leaving them to take on a cloudy consistency along with orange, copper and deep gold colors. These wines will be moderately sparkling to slightly fizzy. In terms of flavors, they can be a little tart and tangy, resembling a bitter beer or dry hard cider or even offer up a kiss of sweetness.

ROSE SPARKLING WINE — This is another reason why the sparkling category is so fascinating because it keeps us on our toes with its different practices. We learned that a rosé wine is made from red skin grapes when making a still, non-sparkling wine. It's different when making a rosé sparkling wine. To make a sparkling rosé wine, the winemaker will, in essence, make two wines – one red wine and one white wine. The white wine will be used as the majority of the base wine, maybe 70 to 90 percent of the white wine. Then the smaller ratio of red wine will be mixed in with the white wine for flavor, and of course, to garner the rosé color. These wines tend to feature a lot of bubbles and personality. They can be Brut, Extra Dry and Demi-Sec. In terms of notes and flavors, they can give off hints of cherry, cranberry and strawberry-rhubarb pie notes.

RED SPARKLING WINE — If red wine seems to be more common place in terms of still wines, they are definitely a smaller population when it comes to sparkling wine. But they do exist. In this case, these are simply still red wines that undergo a form of secondary fermentation to become carbonated. Since these are red wines, there will be a certain level of tannins to them. That tannic structure, typically on the softer side, makes them great with meats, cheeses and dishes with some sort of fatty content. And these wines can be quite fun in the warmer months as you can enjoy a cold glass of sparkling red outside or with your grilled dishes. These wines can range from highly carbonated to slightly effervescent. They can be Brut to Demi-Sec. And in terms of flavor, they can offer up bright cranberry to deep black cherry notes. Specific styles include Brachetto, Pinot Noir, Lambrusco and Shiraz.

: the sparkling wine food pairings

Believe it or not, sparkling wine is the easiest way to make your wine and food pairings a major success. Many consumers enjoy drinking sparkling wine on its own or with deserts, particularly when used to celebrate special occasions.

However, sparking wine can be enjoyed all-year-round and is quite tasty with a variety of foods.

There is a section about food and wine pairing tips in CHAPTER SEVEN: **the shopping, serving and storing**. But since we're here, there's no better time to briefly touch on it a bit with sparkling wine.

The quality that makes sparkling wine ideal for food pairings is the acid content. Remember, all wines have acid. Sparkling wines, however, boast a higher acid content. It can range from moderate- to high-acid, but all levels are great in terms of food and wine pairings.

In conjunction with the acid, the slight nuances in body styles – light, medium and medium-to-full (medium plus) – make sparkling wines extremely versatile. Then you layer the variety of color types and you have a no brainer, win-win pairing situation.

The white, orange and rosé sparkling wines are ideal to enjoy as an aperitif, throughout a multi-course dinner or with desert – depending on the sweetness level.

Red sparkling wines work nicely with antipasti platters, barbecues, grilled vegetables and a variety of proteins. It can also be served after dinner depending on the sweetness level of the wine.

When you have a food and wine pairing dilemma, I always say: "when in doubt, go the sparkling wine route."

CHAPTER FOUR: **the grapes**

If the "place" sets the stage for the overall wine experience, then the grapes are the diverse cast of characters at the heart of the story.

Given the proper podium throughout the major wine regions of the world, these grapes can change into extraordinarily complex entities during their second act – their metamorphosis from grapes to wine. And some crucial guidance and direction by the hand of experts along the way doesn't hurt.

Since we will focus on the unique characteristics of these grapes, the grape varietals will be the main "characters" in this chapter. Like great characters, these principle grapes transform into wine and, in some cases, world-wide sensations that bring a certain something special to the table.

While there might physically be fewer grape varietals in the world than there are SAG (Screen Actor Guild) members in Hollywood, there are still quite a lot of grape varietals to get to know.

But before we get to know our leading cast in this small, "less is more" production, I need to back up and explain a few things.

Let's revisit the term grape varietal first.

Varietal represents the names of specific grapes. In the wine world, the names of varietals come from the Vitis vinifera species of grapes. Some professionals also refer to them as grape "varieties." However, varietal is how it will be referred to in this book.

This Vitis vinifera species is known as the wine grape. Its origins are linked to Mesopotamia – now modern-day Iran, Iraq, Georgia, North Africa, Armenia, Syria and Turkey – as we referenced in CHAPTER ONE**: the history**.

There are thousands of grape varietals from the vinifera species found globally in the world. Yes, that's right. Thousands of grape varietals!

Experts have estimated that there are between 5,000 to 10,000 grape varietals from this species that have developed over the last 8,000-plus years.

If that number sounds strange or excessive, compare that to the number of varietals found in Italy. In that country, alone, it's said to boast up to 3,000 grape varietals. So, the larger number – 5,000 to 10,000 – doesn't sound so far-fetched in the global context.

With that large number of available grapes that can be turned into wines, there is definitely a wine sure to satisfy everyone's personal palate and interests.

You can have the opportunity to sip on everything from the letters A to Z; like an Airen from Spain to a Zinfandel from California and everything in between.

Okay, wait! I know what you're thinking. That's a lot to process: 5,000 to 10,000 grape varietals?

Relax. Take a breath. Take a sip of wine. It'll be fine.

Coinciding with the theme of the book, I opted to focus on a very small population of some of the most widely consumed wine grapes in the world. With that being the case, this chapter will *only* focus on five (5) white skin grapes and five (5) red skin grapes.

I wanted to emphasis the number five for each a category to make the point that this not going to be difficult at all.

This chapter will showcase each of the grape's back stories, how they are perceived by certain audiences and how they show off their diversity from place-to-place.

: the white grapes

The natural order of wine tasting is to go from light wines to heavier wines. Therefore, in this book, we will start with the white wines grapes first and then move on to the red options.

I've chosen five white grapes that I believe most people have tried at some point and / or that are significant in the wine education process.

The grapes are placed in alphabetical order. And the grape and the wine it turns into will sometimes be used interchangeably in this chapter.

Without further ado, ladies and gentlemen, here are the leading white wine grapes in "The Less is More Approach to Wine."

: **chardonnay**

Let's quickly address the elephant in the room in terms of Chardonnay.

Some people are heads-over-heels in love with it. Some people just out-right hate it. Well, hate might be too harsh. Let's say "strongly dislike."

Those who find themselves "strongly disliking" Chardonnay fall into a category we call A.B.C: "Anything But Chardonnay." They won't drink it. They won't touch it. They won't smell it. They won't get close to it.

When I ask my wine classes about their thoughts about Chardonnay, the room is normally split down the middle in terms of their affection, or lack thereof, of the grape.

That's understandable to have such a strong reaction to a grape. But I want to encourage all fans of wine and wine education to keep an open mind as we get to know Chardonnay a little better.

Chardonnay has long been considered an "It" grape for centuries.

Known as the "winemaker's grape," people involved in making wine love the adaptability and flexibility that Chardonnay offers. Part of the reason is that it is a fairly neutral grape, in flavor and aromatics. As a result, it can be shaped into different styles. Plus, it can be grown in just about every wine region in the world – showing up in both cool and warm climates.

In its native country of France, Chardonnay is a noble (prestigious) white grape that reigns supreme. The grape accounts for about 60 percent of plantings in both Burgundy, where it originates from, and also in Champagne.

However, it seems that over the last few decades it has lost much of its following. In my humble opinion, a lot of Chardonnay has fallen victim of typecasting, bad direction and low-budget productions.

For the last thirty years, or so, in the United States, Chardonnay has been cast as either a superbly rich white wine found in America's best steakhouses ranging from Connecticut to California. Or on the converse, it has been regulated to convention centers from the Pacific Northwest to the Dirty South as an inexpensive, unbalanced and sometimes one-dimensional white wine.

One of the key contributors to the polarizing reaction to the grape is its oak influence. The use of oak barrels can be both loved or loathed in Chardonnay.

As we learned in the previous chapter, wine can be fermented in neutral vessels like Stainless-steel tanks, amphora clay pots, or concrete egg tanks. Those vehicles don't add any flavoring or coloring the wine. These vessels allow the elements of the grape to be showcased more front and center.

Cool-climate Chardonnay grapes produce more lean, acidic and mineral driven wines with hints of green apple, lime, olives and flint. The Chardonnay wines made from these regions tend to be produced more in neutral vessels.

Then some Chardonnay grapes have oak interaction during and / or after fermentation. The oak can be neutral oak barrels or barrels treated by fire to impart subtleties of flavors, aromas and textures into the finished product.

In more temperate and warmer climates, the wine showcases rounder notes (especially when interacted with toasted oak): baked apple, lemon curd, pear, pineapple, vanilla, butter, camp fire and sometimes a bit of residual sugar. This is when you have grapes that are riper due to the warmer weather that can stand up against a more powerful oak presence.

Then going back to the topic of low-budget production, some producers tend to opt for oak chips to let soak with the Chardonnay juice to get that oaky flavoring into the wine. You will find this with many of the inexpensive Chardonnay wines on the market.

Oak barrels tend to be fairly expensive. A French Oak barrel can cost between $1,000 and $1,500 and an American Oak barrel can cost between $700 and $1,000 per barrel.

Oak flavor is also a way winemaker can mask any issues they have had in the growing of the grape through the harvest to make the wine drinkable and marketable.

When interacting with my students, I remind them that Chardonnay is still a very classic grape and spans the gamut in production styles. Although it can be typecast, this grape has a lot of range.

As a sommelier and wine educator, I suggest trying a Chardonnay from a different region of the world or one that has a contrasting production style than what you have been accustomed. Then see if that adjusts your opinion about Chardonnay a bit.

Other countries of note in which to explore Chardonnay include Chile, South Africa, Sonoma County, New Zealand and Oregon.

While Chardonnay is indeed one of the stars in this chapter, there are some appropriate understudies that could serve as nice alternatives: Viognier, Marsanne, Roussanne and Falanghina grapes.

: **chenin blanc**

"Chenin what? Chenin who?"

I joke, but that sums up a consistent response when I present people with this noble white grape hailing from the Loire Valley in France.

Chenin Blanc offers up brilliant performance after brilliant performance around the world. However, it suffers from the lack of exposure.

It is a grape varietal often overlooked by many wine stores and restaurants. Therefore, a good deal of wine drinkers may not even know that it exists.

Out of the five principle white grapes featured in this chapter, Chenin Blanc would be the underdog of the group.

Being exposed to Chenin Blanc for the first time can be a little mind blowing. At least it was for me. This grape creates wines with fresh, clean and rich impressions that tend to leave wine lovers a little speechless.

Consumers often tend to search for ways to describe the wine by looking to make comparisons to other grapes varietals. If one was forced to make associations, I would suggest that it can be somewhere between a Chardonnay and a Sauvignon Blanc.

That's a blanket statement I am forced to make at times to give wine drinkers something to wrap their minds around. However, with its eccentric nature and ability to be grown in select parts of the world, Chenin Blanc is in a league all its own.

In terms of character, this grape has dynamic range.

In its native land of the Loire Valley, particularly in a region called Vouvray, the grape can showcase lush hints of lanolin, damp hay, melon and honeysuckle. The grape varietal can be fermented into a dry style, a semi-sweet style or even a sparkling style.

Savennieres, a winemaking region also in the Loire Valley, produces bone-dry Chenin Blanc styles that tend to be sharp, mineral driven and slightly funky, with doses of citrus, dried pear and oak. A little peculiar and off the beaten path, this style can be a bit of an acquired taste. Some of these wines are also made in an orange-wine style.

New-World Chenin Blanc wines usher in a type of freshness, predominately associated with these warmer regions. In South Africa, where it's affectionally referred to as Steen, there are zippy notes of lime on top of honeydew on top of hints of crushed peach.

The Chenin Blanc styles from California regularly offer up the same playful and youthful styles of South Africa. But given the huge influence of France on the wine world, some of the wines made from this varietal in California throwback to some of the clean, crisp, earthy, old-world styles found in Loire.

Chenin Blanc can be eclectic, but also brilliant, unexpected and evocative.

While exploring the varied Chenin Blanc styles, other lovely grape alternatives to consider would be Grenache blanc / Garnacha blanca, Muller-Thurgau, Semillon and Viognier.

: **pinot gris / pinot grigio**

Pinot Grigio is a fan favorite. There's no question about that. As a wine, it has an unparalleled way of showing off its soft approachability and a simplicity people have come to love.

However, as a grape, Pinot Grigio is more than meets the eye. Literally!

Let's first start with the name. In the header of this section, you might have noticed two names instead of the one grape varietal name found in other sections. That's new, right! And it has probably already led to some raised eyebrows, questions and confusion.

So here is the deal. Pinot Gris and Pinot Grigio are the same grape. It just goes by two different names. Like I might go by Charles to some and Chuck to others. But there's just the one me. Two names, one entity.

Pinot Gris is the French name for this grape. Pinot Grigio is the Italian name for the grape.

More recognizable by its Italian name, the grape has roots in France – the Alsace region to be specific. Pinot Grigio, however, is the name that is most widely used by consumers.

Still sticking with the name, we want to go to the second part of the name: the gris and the grigio.

Gris means grey in French while grigio means grey in Italian.

That is important because when you see this grape on the vine – in any part of the world in which it grows – you will see skins of green, grey, pink, purple and yellow. But despite the range of colors on the skins, the grape falls into the "white grape" category because the skins are commonly removed before the fermentation process.

Here is another reference to orange wines. Pinot Gris / Pinot Grigio is a lovely grape in which to leave the skins in contact with the juice while fermenting. Winemakers can extract soft to deep colors from the skin, turning the wine into orange, copper or even intense dark-pink hues.

See what I mean. This grape is definitely way more than meets the eye.

On the palate, as a white wine, it can also offer up flavors and textures out of the general scope of what people associate with the grape. So, while technically the names can be used interchangeably, there are some important stylistic notes that separate the Gris from the Grigio.

In France, the grapes are transformed into wines that can have some lovely texture, depth and weight. It's a dry, sunny environment that helps produce aromatic and flavor notes including fresh lemon, pear, flint and flowers. Oak aging in some cases allow a richer texture to take center stage while the acidity is more subdued. The Pinot Gris from this region can be vinified to be a dry, off-dry or sweet.

The Pinot Gris wines from Oregon winemakers can also follow this same French style of winemaking.

But when you jump over to Northeastern Italy – especially in Veneto – you get the style of Pinot Grigio in which most people are familiar: simple, soft, approachable.

That has been the consistent, even-keel performance of the grape which has been attributed to the Pinot Grigio from this portion of Italy. Light- to pale-yellow in color, this Pinot Grigio lures you in with hints of citrus and stone fruit, wet rocks, almond skin and moderate alcohol.

Given the right circumstances, however, Pinot Grigio has layers and depth that can leave people craving for more. When the grapes come from Veneto's neighboring regions, like Friuli and the Alto Adige, you get wines that are distinctive, spirited and precise. The clean, fresh, mineral style is celebrated in those regions. The texture deepens a bit when winemakers in Collio, Italy, will play around with skin contact to produce copper toned, silky smooth versions featuring pear, banana, jasmine and orange blossom notes.

To expand your interests out of Pinot Gris / Pinot Grigio a bit, other grapes that might interest you are Muscadet Sevre et Maine, Pinot Blanc, Vermentino and Picpoul de Pinet.

: riesling

Into every life a little Riesling should fall. That is at least what most sommeliers, like me, believe.

Yet it is not easy trying to convince the masses of that.

It's no mystery there. We all know why Riesling can be a hard sell for many wine drinkers.

People associate this grape with what? Lots of sugar!

It is major victim of typecasting, beholden to the stereotype that Rieslings are *all* sweet.

That notion really is heartbreaking.

If there is one takeaway, we learned so far from this book, it is that the wine world is full of surprises.

The major surprise when it comes to Rieslings, is that most Rieslings are actually dry.

Yes, dry. That is not a typo or a misprint.

The majority of Rieslings produced around the world fall into a dry or off-dry category.

Recapping on what we learned in the previous chapter, "dry" means there is little, or no, sugar left in the wine after its fermented. Off-dry means there is a small amount of residual sugar left in the wine.

But yes, I understand why that "sweet" perception exists.

There is a smaller population of Riesling wines that can be quite sweet: semi-sweet or even dessert wine in sweetness.

A generous dose of residual sugar from wine landing on a palate can be quite jarring. That is especially true if your palate is accustomed to dry wines. Or if you generally have a stronger preference for dry wine over sweet wines.

With a little or a lot of residual sugar, one might think a wine made from this grape would not have a regular place at the dinner table. And he or she might not be highly motivated to order it at the bar after a long day's work.

But surprise, surprise. Riesling is a perfect vehicle for both situations.

Riesling is ideal to pair with a wide range of cuisines – from the native German fare like Bratwurst, sauerkraut and potatoes to Thai dishes like spicy Pad Krapow Moo Saap (Fried Basil and Pork). And the high-acid structure of Riesling makes them fun to drink and very refreshing after a long day of work.

Native to Germany, grown in regions like Mosel, Nahe, Rheingau and Rheinhessen, this grape has no choice but to develop in cool-climate conditions.

Pushing right up to the 50th degree latitude range, Germany is one of the coldest traditional winemaking regions in the world. The proper soil, along with heat retention capabilities, is crucial for proper ripening, viticulture, viniculture.

Those conditions also allow Riesling to capture the expressive terrior of its region. That makes it one of the most appropriate grapes to show off a "sense of place" in a glass.

With the challenging growing conditions, Riesling can show us how ideally balanced wine can be. The high acid content lifts the residual sugar, providing light qualities on the palate and minimizing any cloyingly sweet characteristics associated with wine containing substantial amounts of unfermented sugar.

The sugar levels of German Rieslings can be easily understood by the verbiage on the labels. Dry styles will have the name **Kabinett** which indicates the minimal ripeness levels by the government. These will be the dry styles, with the word **Trocken**, also featured on the label meaning dry to little residual sugar in the wine.

For more fruity styles, opt for the **Spatlese** ripening level from Germany, which will have more detectable sweet characteristics. These will be more on the off-dry style, a little more residual sugar than the dry styles.

Then for a semi-sweet style, the **Auslese** ripening level will be ideal. These are grapes picked later in the harvest to allow for the optimal ripening levels.

A great way to determine the possible level of residual sugar on Rieslings in Germany, and around the world, is to look at the alcohol content on the label. Wines with 10 percent ABV, alcohol by volume, or less tend to have more residual sugar and will have more perceived sweetness. Wines with 11.5 percent or more ABV will typically be on the drier side.

The aromatics on Rieslings can be quite interesting too. You can smell anything from lemon, lime, apple, stone fruit, papaya and an unforgettable aroma of petrol, like a brand-new shower curtain liner.

Classic dry and off-dry selections can be found in Germany and Alsace, France. In the Pacific Northwest, Riesling tends to fall into the dry realm in Oregon and the off-dry style in Washington State. In Upstate New York, the Finger Lakes region produces some rich, off-dry styles that are lush and supple. And in Australia, the Rieslings from areas like the Barossa Valley are bright, crisp, zesty and vibrant.

Richer styles and dessert Rieslings can have quite the shelf life, sometimes with the ability to age 20 to 30 years.

Other grapes of interest that could be considered similar to Riesling are Torrontes, Gewurztraminer and Moscato.

: sauvignon blanc

I would be hard pressed to meet anyone of legal drinking age that has not had some type of exposure to the Sauvignon Blanc grape.

It's like the mega star of white wine grapes, having gained a huge following due to its youthful nature, versatility and lively personality. And it has been enjoying that spotlight for some time now.

In terms of performance, Sauvignon Blanc can range from the subtle to the dramatic. It doesn't give you a very deep or weighted performance that makes you wax poetic about the grape.

It's not that type of grape at all. Its strength lies mostly in evoking an intense first impression that stays on your mind and on your palate for a while.

While many consumers are loving the Sauvignon Blanc from New Zealand, this grape also stems (no pun intended) from France. More specifically, the Bordeaux region. In Bordeaux, it is usually blended with a grape called Semillon. However, sometimes you can find single-varietal Sauvignon Blanc wines from Bordeaux.

Sauvignon Blanc from Bordeaux produces wines that are medium bodied, offering rich notes of citrus and pineapple and apple.

This grape is also grown successfully throughout locations in the Loire Valley of France. In regions like Touraine, Quincy, Pouilly-Fumé and Sancerre, this grape is *the* white grape of these areas.

With a core of strong citrus notes, ranging from lemon, lime and grapefruit, the Sauvignon Blanc grape produces racy and electrifying performances on the palate. Other notes include herbs, lemongrass, bell pepper and clean minerality.

The minerals in Loire that are imparted into the grapes range from limestone, chalk, clay and flint. Areas with a higher level of flint rocks in the soil, like Pouilly-Fumé, tend to offer up a smoky expression.

Oak is usually bypassed in the production of these wines, although some winemakers, like in Napa and the Loire, might opt to use some oak for unique stylistic purposes.

In New Zealand, where the Sauvignon Blanc grape is having a serious moment, those wines stand out among consumers due to their strong aromatics and bright flavors. Here you might get a strong dose of grapefruit, gooseberries, stone fruit (peaches, plums, apricots), passion fruit, mango and kiwi. It's almost like having a tropical fruit salad sprinkled with freshly squeezed pink grapefruit juice.

The styles can vary a lot in warmer places like Chile, California and South Africa. In sunny California, you get the riper stone fruit elements, hints of green apples and a little touch of tangerine. The wines will have a slightly richer texture. And some wines will be aged in oak for a while to present a lemon-butter flavor with a slightly smoky effect.

In Chile, Sauvignon Blanc brings to mind much of its citrus quality with a touch of lemongrass. While South African Sauvignon Blanc grapes present a strong showing of fresh citrus, stone fruit notes and chalky minerality.

As distinctive as Sauvignon Blanc is in the wine world, other grapes like Albarino, Verdejo, Viura and Gruner Veltliner are great and equally exciting alternatives.

: the red grapes

Red grapes tend to have a special place in the hearts and minds of wine lovers. The wine from red-skin grapes can signify comfort, sophistication, luxury and prestige.

It can bring many scenarios to mind; like thoughts of warming up after coming in from the cold, eating a big, succulent steak or decompressing after a long, hard work day like the fictional character Olivia Pope on the television show "Scandal."

It feels that wine made from red grapes need little to no introduction. But for education purposes, I wanted to properly introduce these grapes (or in some cases re-introduce these grapes) in this section.

For the reds in this chapter, we have five well-known and, for the most part, well-respected grape varietals: Cabernet Sauvignon, Malbec, Merlot, Pinot Noir and Syrah / Shiraz.

Everyone has his or her own opinion about these grapes. Personal critiques and preferences aside, hopefully showing these grapes in a different light will help develop an overall new respect for these wine grapes and maybe even a deeper appreciation.

: cabernet sauvignon

Classic. Relevant. Timeless.

Cabernet Sauvignon is one of the most regal, noble and respected red grapes on the planet.

A true superstar, Cabernet Sauvignon comes to mind when people think of red wine. That admiration seems to be widely unchallenged in the wine world from yesterday, today and, most likely, tomorrow.

This grape is nearly everywhere you want to be like an American Express credit card. It's in steakhouses, happy hours, corporate events, barbecues, private dinners. You name it, Cabernet Sauvignon is most likely there.

In terms of plantings, it can be found in many wine regions around the world since it is fairly adaptable – when grown in the right setting, of course.

If I was forced to pick one word to describe why people love Cabernet Sauvignon so much, I would choose the word "consistent."

Despite where it grows, the wine from the Cabernet Sauvignon grape stays true to its innate character, making it reliable and recognizable to consumers.

It is like the "Gold Standard" in which most of the red wine grapes are measured. But like any true Superstar, Cabernet Sauvignon will offer up some unexpected variations in its performances.

A native of Bordeaux, France, Cabernet Sauvignon is one of the most noble grapes in that region. However, you'd be hard pressed (no pun intended) to find many notable, high-end wines that feature Cabernet Sauvignon exclusively in the bottle from this region.

Known for using multiple grapes to blend together for a signature wine, Bordeaux blends Cabernet Sauvignon with Cabernet Franc, Petit Verdot, Malbec and Merlot. Cabernet Sauvignon and Merlot blends tend to be the most common blends from this region.

While Cabernet Sauvignon is not the most planted grape in Bordeaux region, it does, however, offer up a lot of power in terms of structure and fruit.

On the Left Bank of Bordeaux, Cabernet Sauvignon ripens better in this slightly warmer region which creates juice that has more tannins, color, alcohol and moderate acidity. In terms of aromatics, it presents notes of black fruit, blackcurrant, licorice and mint.

Cabernet Sauvignon successfully navigates the leveled landscapes of wine by being the "People's Choice" for their Monday night wine to a "Critics Choice" for some of the most publicized Bordeaux Blends.

It is said that imitation is the highest form of flattery. That is very true of other winemakers around the world when it comes to Cabernet Sauvignon-based Bordeaux-style blends. Any winemaker, in notable regions, that grows Cabernet Sauvignon will try their hand at creating their own take on a "Bordeaux Blend."

And in warm weather areas, Cabernet Sauvignon does amazingly well on its own as a single varietal wine.

In Napa Valley, the Cabernet Sauvignon grape is like liquid gold. The region got the world's attention during the Tasting of Paris in 1976 when a Napa Valley Cabernet Sauvignon beat out several of the French Bordeaux wines in a Blind Tasting. Napa Cabs became all the rage.

Down in the Southern Hemisphere, Australia has been able to produce everything from high-end, award-winning Cabernet Sauvignon wines to juicy everyday table wines. The country has also had great success blending Shiraz and Cabernet Sauvignon together as a sort of spicy spin on Bordeaux-style blends.

In South America, you can find plantings of Cabernet Sauvignon in many areas throughout the continent. But the primary producers are found in Chile and Argentina. Both countries have their signature red grapes they are now known for, yet Cabernet Sauvignon still represents a nice amount of the reds produced in those regions.

And then we have South Africa. The Cabernet Sauvignon wines from South Africa are almost like a meal and a beverage in one. There is a generous dose of "gamey," smoked-meat quality represented in many of the Cabs from this region. That is in addition to the signature black berry, black currant, blue berry and "jammy" notes that the grape is known to produce.

If you are looking to enjoy other wines that are similar to Cabernet Sauvignon, try Tinto Fino (Ribera del Duero, Spain), Tempranillo (Rioja, Spain), Touriga Nacional (Portugal), Nero D'Avola and Primitivo (Southern Italy).

: **malbec**

Malbec is one of the most invigorating and exciting things – in terms of wine – to have come out of South America.

But most people are stunned to learn that the Malbec grape has deep roots in the South of France. The grape was introduced to Argentina in the mid 1800s by the French.

Yes! I know! There goes France jumping into the picture again. It's inevitable. But we'll get back to France a little later.

Fans of red wine have fallen hard for Malbec over recent years, specifically the ones hailing from Argentina.

In the specific terrior of Argentina, the Malbec grape enjoys a super unique environment in which to develop. The environment helps foster the ultimate ripeness of the grapes, resulting in a bold, spicy and smooth wine. These characteristics tend to be the ones most sought after by consumers.

Deep fruit notes like plum, black cherry, black currant and blueberry are plentiful in these styles. Then with the richer texture and structure of the wine, Malbec spends time in oak barrels which help transmit spice, tobacco, umami characteristics and Earth.

However, it's the elevation in the foothills of the East Andes mountains that also helps Argentina cultivate such a distinct style. There are some plantings in the low-lying areas, but many of the winemakers are using grapes that are planted on vines as high as 2,500 to 5,000 or more feet above sea level. That's a huge contrast in terms of the environment and terrior of France.

This ancient yet sophisticated high-elevation farming was developed on the Andes Mountains by the indigenous Inca people around the 15th Century. Out of necessity, it allowed for a wide range of crops to grow properly under difficult circumstances: the cold mountain climates, desert conditions and humid jungle conditions that surround the region.

Argentine Malbec started to become an international sensation over the last 15-plus years. But many of its fans had, or have, no idea it is originally from France in the regions of Bordeaux and Cahors.

In Bordeaux, the Malbec grape is typically blended with the other red grapes from that region (Cabernet Franc, Cabernet Sauvignon, Merlot, Petit Verdot) to make up the classic red blend.

However, in Cahors where the grape also goes by the name Cot, the grape mostly stands alone as a bold, Earthy wine with concentrations of aged fruit and hints of tar. This is a very different style than the Malbec wines of Argentina.

Besides Argentina and France, Malbec is also grown in Chile, Italy, Spain, South Africa, New Zealand and the United States, but in much smaller quantities.

In Chile, the next-door neighbor to Argentina, single varietal Malbec has a more subtle style from the cooler regions in the country.

California winemakers will often use Malbec as a blending grape, used in many of the fashionable "Bordeaux-style Blends" coming from Napa Valley and the North Coast.

Other grapes to try, if you enjoy Malbec, are Garnacha, Petit Verdot and Nero D'Avola.

: merlot

Merlot is as beautiful as it is misunderstood. It as elegant as it is strong. And it is as famous as it is ignored.

Originally from France, Merlot commands the love and admiration of Right Bank Bordeaux wine lovers and critics alike.

Merlot wines from that region of Bordeaux can be some of the most expensive wines in the world. A bottle of Chateau Petrus can cost you on average $2,600.

That's just one bottle of wine. One standard 750ML sized bottle. Not a 3-liter or 6-liter bottle.

For those of us that fall into a lower tax bracket, lovely Merlot options can also cost consumers just under $10 in some cases. Or the sweet spot can cost anywhere between $15-$35 dollars from certain regions.

The versatility of Merlot is extraordinary. It has a lot to offer and has a wide mass appeal. However, in the United States and some other parts of the world, it gets no respect – or love for that matter – these days.

Merlot has been unfairly and unjustly tossed aside as a "has been" grape by a lot of wine drinkers.

Blame it either on the much-documented "Sideways" effect of the movie or the grape just getting lost in the shuffle of "lesser known" grape varietals hitting the global market place.

But this regal, noble grape definitely doesn't get the prestige and attention it was accustomed to receiving on the world stage.

That is a real shame. People who choose to opt out of Merlot wines from around the world are missing out on the most attractive hints of plush plum, mocha, black cherry, walnut and tobacco with variations of sage, rhubarb, eucalyptus and mushroom. That all, of course, depends on the terrior and winemaking practices.

As life has proven to be very cyclical, don't be surprised if Merlot makes a major comeback in the near future.

Try other grapes like Carménère, Grenache and Graciano, if you want some wonderful alternative options to Merlot.

: pinot noir

The allure of Pinot Noir is undeniable.

From its striking appearance to its distinctive nose and its graceful mouthfeel, Pinot Noir has that certain "*Je ne sais quoi*" (*I don't know what*) quality.

Unfortunately, behind the scenes – from grape to glass – the process isn't so lovely. It actually can be quite stressful for winemakers.

Pinot Noir is very finicky and difficult to grow. With all the challenges it presents to winemakers, the grape can be perceived as somewhat of a "Diva."

It has affectionately been coined the "Heartbreak Grape."

One of the main reasons the grape can prove challenging for winemakers is because it needs a long, cool maturation process in order to grow properly. It ripens early, therefore the grape needs a suitable climate to allow the structures to develop fully over time. It also has thin skins and grows in very tight, large clusters, making it susceptible to rot and mildew.

When things go wrong with the grape, they go *very* wrong. On one hand, the grape can result in a very green, under-ripe style of wine. Or the opposite can happen. It can offer up a hot, overripe wine with cooked qualities.

All the stars have to align properly for Pinot Noir to offer up a balanced performance.

However, when planted in the right regions and managed properly in the vineyard, it shows off a very expressive terrior. That makes it one of the most fun red wines to drink on the planet.

Another native of Burgundy, France, Pinot Noir is *the* major, noble red grape of that region. It produces some of the world's most sought after and collected wines. Wines from this region tend to be light bodied in style, offering up hints of raspberry, cherry, cranberry, wet leaves, pepper and flinty minerality.

High-end options like Domaine de la Romanée-Conti (DRC) wines can easily cost you $1,000-plus per bottle retail for recent vintages. The wines sometimes go up to $20,000, or more, per bottle at auction for much older vintages. It's insane.

But most wine lovers enjoy Pinot Noir options from anywhere between $20 to $100, give or take some coins depending on the style and region you select.

Surprisingly enough, Germany produces some beautifully bright, Earthy Pinot Noir wines that go by the name Spatburgunder. The same can be said of these grapes grown and produced in Oregon in the Pacific Northwest of the United States.

In Northeastern Italy, the Pinot Noir is called Pinot Nero, nero meaning "black" in Italian like noir in French means black. Characteristics of wines from these cool-climate regions, include notes of roses, spiced strawberry and mushroom.

In warmer climates like Argentina, Chile and California, Pinot Noir gets a little more sun that puts the wines on the cusp of light-to-medium bodied wines in terms of weight. Grown on higher elevations, these wines pronounce notes like roasted beets, plum, suede and Asian five spice seasoning.

While Pinot Noir is definitely one of a kind, other grape options can be Gamay, Blaufrankisch, Zweigelt and Corvina.

: **syrah / shiraz**

Bold, burly and robust with a brooding appearance, Syrah's massive presence is often tempered by its lush approachability.

It's like a big ole Teddy Bear you want to cuddle up with.

Syrah is a supple, spicy red wine that is so dynamic that it regularly goes by two names.

As you might have guessed, Syrah is the French name. That name is used for grapes found in various regions of France, Spain, Italy, North America and South America.

Shiraz is an Australian take on the name Syrah. With Shiraz being so catchy and popular, some winemakers in South Africa have come to use that name as opposed to the French name.

Comparing Syrah from France and Shiraz from Australia is a compelling, nearly cut and dry study into Old World versus New World terrior and winemaking practices. It can show off the differences of "Old World" and "New World" quite quickly.

As Syrah – in its native Rhone Valley in France – it can be medium-bodied, herbaceous, spicy, meaty and Earthy. In the Northern Rhone Valley, the Syrah grape is made into a single varietal wine. However, at times, winemakers will add a small percentage of a white wine grape called Viognier for aromatics. Notes can range from prune, eucalyptus, blackberries, violet, smoked meat and potted plants.

Blends of Syrah abound in the Southern Rhone Valley and other parts in the South of France where the grape is blended with Grenache, Cinsaut, Carignan and Mourvedre. But in the region of Chateauneuf-du-Pape (The New Castle of the Pope), Syrah can be blended with up to 12 other grape varietals to make for a highly distinctive wine. Wines from these southern regions in France tend to be richer, denser and sometimes more fruit-driven.

As Shiraz – grown in Australia – the grape takes on rich, ripe, succulent, spicy and generous fruit elements. Notes can range from blackcurrant, ripe plum, camp fire and mixed-berry jam.

The grape has also found a home in other Old-World regions like Italy and Spain. And in the New World, it is grown in Argentina, Chile, Greece, South Africa, California and Washington State in the United States.

Syrah / Shiraz from around the world tend to feature moderate to firm tannins, however the ripeness levels of the grapes can often soften the mouthfeel of the tannins on the palate.

And many winemakers produce Southern Rhone Valley blends in their respective regions around the world in the form of GSM (Grenache, Syrah, Mourvedre) blends.

You might also want to explore Petite Sirah, Aglianico and Cannonau for grapes comparable to Syrah/Shiraz.

CHAPTER FIVE: **the tasting approach**

You've been patient.

Or maybe you weren't so patient. Maybe you cut straight to the chase.

I know my wine people – both spectrums from the studious to the impetuous. Whatever approach you take with this book, and with wine, it is very personal and individual.

I'm just glad you are here. Now that you are here, this is the point where the real magic happens from a personal perspective.

It is officially time to learn how to *taste* wine.

But not just casually sip on wine like you have done several dozen times before. We are going to purposefully engage all of our senses to get the full spectrum of nuances from wine.

Don't worry. This will be fun and easy.

Tasting comes naturally. Scientists say our "tastes" first start to develop in the womb. Then we continue to learn, as we grow, what we like or don't like. Those preferences are shaped by our experiences.

In fact, studies suggest that we develop our tasting phobias between the ages of two and ten. That's quite young.

And as babies – and sometimes adult babies – we immediately spit out what does not quite suit our personal tastes.

Taste, whether it's food or wine, is actually a combination of several mixed sensations. In the case of wine tasting, this is more of a formal exercise that eventually turns into a habitual practice.

When I teach wine classes, I take my students through a very sensual approach to tasting. As a result, we'll engage our senses: sight, smell and taste. Then we'll file those notes away in our memory for future reference.

To me, these are important steps that are central to understanding wine in a real, tangible way. These stages help us discern the wine's quality, value and pedigree. It also allows the taster to form his or her own opinion about the wine.

It is natural to decide immediately if you like it or not. However, developing your own personal wine palate is not organic to most people. Besides coming to the conclusion of whether or not we like the wine, we should also determine "why" we like it or don't like it.

Think about it this way. Think about wine as food as you consider the upcoming tasting exercise. We've had a lifetime of trial and error, tasting and sampling and smiling and frowning over food options.

We know what we love. We know what we don't love so much. We know what we can tolerate. And we most certainly know what takes us to a very special "happy" place.

When it comes to wine, most of us in the United States haven't had a lifetime of trial and error: tasting and sampling and smiling and frowning over wine. Most of us were only able to access wine in our late teens or early 20s for the first time.

Therefore, it is very easy for us to feel lost when it comes to enjoying wine or even developing an appreciation for it. The later a person gets exposed to certain flavors, the more difficult it might be to develop a "taste" for those items.

It might be challenging, but it is not impossible.

While you are not between those crucial ages of two and ten years old anymore, we can still train our brain to like new things. It just takes some time and practice. And there's no better time to start building that knowledge than right now.

Trust me. There are much worst things in life than going through the exercise of tasting and sampling wine to see what's ideal or not. Am I right? Or am I right?

So how do we begin to learn to taste wine?

In this chapter, we will begin with The Five "S's" of wine tasting. You might have heard about these. There are several variations used by wine experts and wine educators. Some educators use four. Some people use six. I like to use five.

The Five "S's" of wine tasting used in this book are: See, Swirl, Sniff, Sip and Savor. During these steps, you will utilize several senses at one time. However, you will focus on one sense per stage during this sensory analysis.

The overarching mission is to educate yourself on how to properly taste wine for the full extraction of flavor and overall enjoyment of wine.

I've taught hundreds of wine classes in New York City over the last nine years. The easiest way for me to get my students to really connect with wine is to walk them through this five-step process. That way they are engaged with what they are personally experiencing in the glass.

Since this is a physical activity, if you can, please grab a glass of your favorite wine and go through each step with that wine.

Let's get started.

: See

This is an important first step that allows you to state the obvious. You can immediately tell if the wine is red, white, rosé, orange or if it has bubbles in it.

Now it's time to uncover some of the not-so-obvious things about the coloring and style of your wine.

Look at the wine. I mean, really look at it. Pick it up and look at the glass of wine in the light. Find a piece of white paper and tilt the glass on a 45 degree or so angle. Take an aerial view, looking down at the wine with the paper below it.

What color do you see? Is it light or dark? Does it have a pale-yellow color or is it more golden for a white wine? For rosé, is it more rose water or deep salmon? For red, do you see bright hints of reds, blues or purples or is it more reddish brown, brick, muddy or tawny in color?

What about the clarity of the wine? It is clear or cloudy? Is there a certain brightness that's evident when looking at the wine? Does it have a shine or glossy appearance? Or is it cloudy? Does it appear to have suspended particles in the wine, making it look dull or textured.

Is this a young wine or can you see some age on this wine? As a white wine ages, it picks up color through oxidation. Think about how an apple, pear or avocado will brown after being cut open. Over time a white wine will turn golden to brownish if it ages long enough. A red wine will lose its bright colors and also start to brown over time. A brownish rim on red wine, when looked at on a 45-degree angle, can show off the age of an older age.

Is this fermented in a neutral vessel or does it have an oak influence? Oak influence on some wines can also affect the color, particularly with white wine.

You notice there are particles in this wine. Maybe it's a flaw? Maybe it's a natural wine? Maybe it's a bit older and has sediment? Maybe it's a white wine that has tartrates?

Does the wine appear to be thin in the glass or have a thicker texture when it moves around the glass? What about those legs?

You might also have come to question if you might actually like the wine based on the look of it.

As you can already "see," there are lots of initial questions that can pop up when analyzing the wine by its appearance.

But the appearance only gives you part of the story.

Let's continue.

: Swirl

The second "S" involves getting more physically involved in the process. We'll need to grab that glass and actively swirl the wine.

Ideally, you have poured yourself a 1- or 2-ounce sized pour. Or maybe you opted for a full 4- to 5-ounce glass. Please don't pour yourself an 8-ounce pour in an 8-ounce glass. We will have some issues with this step if you do.

I know you might love your wine and want to fill that glass right on up to the rim. But wine needs oxygen like we do. So, you want to leave a nice amount of room in the glass to let the wine breathe within the glass.

Take your wine glass and begin to swirl it. You can either keep the base of the wine glass on the table and make small circles. Or, if you feel comfortable, you can hold the glass in your hand off of the surface.

Hold the stem of the glass near the bottom of the bulb and make those small circles. Try to refrain from holding the bulb of the glass, if possible. Vigorously swirl the wine around but be careful not to spill. Continue swirling for a few seconds.

You'll see this juice glide around the glass very elegantly. It might feel a little awkward at first. The more practice you get, however, the smoother the swirling will look.

Swirling the wine in the glass does a few things to set up the next step in the process. First, it allows the wine to interact with oxygen in a more robust way. When the wine rests in the glass, only the top layer of the wine is directly integrating with oxygen. When you swirl the wine, oxygen interacts with the entire wine pour to let it "breathe" or "aerate."

You also coat the glass surface when swirling the wine. As the wine interacts with more of the glass, it leaves a thin layer of wine. That thin layer that coats the glass surface starts to drip down to the wine. When that happens, the alcohol also starts to evaporate. That thin layer then connects with oxygen, which helps release more aroma molecules into the air as the wine evaporates.

This exercise allows the taster to get a better sense of the intricate notes the wine is offering up by concentrating the aroma molecules on the glass.

Another great thing swirling does, especially for white and rosé wines, is that it helps warm up the wine.

Previously, I mentioned that you should hold the glass on the stem by the base of the bulb. Many have been told to hold the wine glass by the bulb to help warm the wine. That would be false and unnecessary.

While that might work for powerful, fragrant spirits like Scotch, Bourbon or Cognac, that doesn't work for wine. Your body temperature will warm up the wine too much, even for red wines. Plus, it leaves tons of finger prints on the glass. That makes it visually unattractive the more the bulb is touched.

But let's get back to the warming up of white and rosé wines. We have a tendency in the United States, and some other places around the world, to drink our white and rosé wine too cold. If the wine is too cold, you can't effectively smell or taste the wine. The main element you are getting out of a very cold glass of wine is mostly the temperature. With a cold temperature, the smell and flavors are tightly locked and difficult to decipher.

So, be sure to have a chilled glass of wine – not too warm and not too cold – for the tasting exercise. More specific information on serving temperatures are addressed in CHAPTER SEVEN: **the shopping, serving and storing**.

At this point, we have put in the work of swirling. Now we get to reap the benefits of that process through sniffing the wine.

: Sniff

The third "S" is all about sticking your nose deep into the glass and taking a series of quick sniffs. This is when you want to see if it is possible to pick out the smells that are present in the glass of wine.

Then try verbalizing what you are experiencing. Focus on what you are *personally* experiencing. Try not to focus on what you think you should say or what someone expects you to say. Pull from your personal smell memory. Think about the items you have smelled and / or tasted throughout your life.

Aromas can connect with past memories. They can evoke emotion, pleasure or discomfort. Work on building associations to what you smell now. Use your current vocabulary of descriptors from your smell and taste memory.

Let the wine take you on an olfactory journey!

What do you smell? It is attractive or does it repel you?

Believe it or not, you can smell everything from apricot to wet dog in wine – and everything in between. Seriously! So, don't be shy to verbalize what you *actually* smell.

What do you smell? Fruit? Non-fruit elements? Flowers? Earth? Wood? Spice? Alcohol? Vinegar? Jam? Wet mildewed towels? Shower curtain?

If you think these descriptors are total BS items that wine professionals just randomly pull out of our nether regions, let's look at the scientific evidence. Refer back to the Aroma Wheel mentioned in CHAPTER THREE that was created by Dr. Ann C. Noble. Google the Aroma Wheel online. You'll see all the various scents that can be associated with wine all match specific chemical compounds

Sniff out clues about the wine. What region might it have come from? In what type of climate was the grape, or grapes, grown? How was it fermented? What type of vessel?

The sense of smell is very crucial to wine tasting, even before we physically taste the wine. Smelling the wine is essentially pre-tasting the wine with our noses. We do most of our "tasting" through our sense of smell.

Scientists conclude that about 80 to 85 percent of what we taste is actually through our sense of smell. It is also said, in the scientific community, that people can smell thousands of different aromas. But we only taste five distinct categories on our palate: sweet, bitter, saltiness, sour and umami.

Does that sound odd? Think back to when you were sick and suffering from a "stuffy" nose. When your nose is incapacitated, you can't really smell your food. As a result, you have a difficult time tasting your food.

That is one reason why our parents, grandparents or caregivers would suggest that we "plug our noses" when taking over-the-counter cough syrup or a homeopathic remedy like cod liver oil. That makes it difficult to taste those unappetizing flavors.

With millions of olfactory neurons in our nose, each person will have their own perception of smells. It really depends on our genetics, cultures and experiences. Some smells might be pleasant to some and offensive to others.

But it's interesting to challenge yourself to smell wine on a regular basis. And then smell with a group of people to see how everyone perceives the various aromas. It truly is a fascinating experience.

If you have a difficult time with smell, work to rediscover your sense of smell. That means smell everything. Smell everything at the farmer's market, the grocery store, the spice market, the flower shop, the bakery.

Smelling wine, and then cross referencing it with items on the Aroma Wheel, can better connect the dots to what you are really getting in the glass.

Smelling and tasting will coincide with each other more as we move on to the next step.

: **Sip**

The fourth "S" is finally about getting that juice into your mouth. This step is all about tasting the wine's unique characteristics and getting a full scope of the wine's "taste." That means everything from the flavors to the textures to the temperatures.

Really dig into the wine during this process. Rather than "drink" as you would normally consume your favorite wine, take your time to really *taste* it.

You want to allow the wine to fully interact with your entire mouth. That allows your taste buds to be engulfed, getting a deeper connection to all the elements.

Take an initial sip of your wine.

When you normally drink wine, it goes straight back on your tongue and then you swallow. Those sips hit your front palate which is called "the attack." You get a first impression of flavors there. Then it hits your mid palate where the flavors can intensify or start to evolve. Then you swallow the wine and that usually gives you a set of unique flavors not picked up from the previous two stages.

After that, the drinker tends to arrive at his or her initial reaction. They like it. They don't like it. They are on the fence. They are perplexed.

However, when you "taste" wine for educational purposes, you want to make sure the wine interacts with your entire mouth.

To get the full picture of what the wine is presenting, you want it to hit the thousands of taste bud in your mouth. In this regard, you are not drinking to formulate a personal opinion about the wine. You are analyzing the wine to fully understand it and make connections to style, typicity, character, balance and regional influences. Then at the end, you can form a subjective opinion about the wine. That means, you can determine if this wine is right for you or not.

There are a few ways you can do this. We'll go from the simplest technique to the most complex approach.

Here is the simple approach. Take a small sip of the wine. Swish it around your mouth for about five seconds like water or mouthwash and then swallow. You should get a rounder, fuller tasting experience as the wine interacts with your entire mouth.

To really magnify that experience, however, you should try the "slurping" technique as I like to refer to it, also known as trilling. This process really helps you get an explosive expression of the wine on your palate.

This process is a bit cumbersome as it has several interlinking steps. It can also take up to 15-30 seconds to complete. So, I'll break them down one-by-one:

1. Pick up the glass and slurp the wine into your mouth. It might sound really rude and crude to do, but you want to bring a lot of oxygen into your mouth along with the wine.

2. While the wine is in your mouth, purse your lips as if you are going to whistle. But instead, you breathe in air. This aerates the wine in your mouth, allowing you to smell the wine through our retro nasal passages (airways that connect the nose and the mouth) found in the back of your throat.

3. Next, you want to swirl the wine around your mouth, like you would do with water or mouthwash after brushing your teeth. Do that for about 5 to 10 seconds.

4. Then swallow the wine.

After completing this process, the wine should taste very differently from the very initial sip you took.

Can you articulate what your palate is experiencing?

How is the acid in the wine? Can you feel your mouth water or salivate? Do you feel the need to swallow a lot more? Or is the acid fairly low or mellow.

What is the wine offering up in terms of flavor? Can you taste a lot of generous, ripe fruit? Is there sweetness there? Or is the fruit more subtle? Is the wine tart or sour? Do you taste non-fruit elements like wood, spice, rocks or metals?

What's the texture like? Is it pretty thin on the palate or does it feel rich and creamy?

All that information was a lot to swallow, I know. It is understandable that many will have a difficult time, initially, with deciphering the aromas in the wine. Sometimes they come easier as larger associations rather than individual scents.

You can start with the larger perspective and work your way down to specific elements. Let's say a wine smells like a Banana Nut Muffin.

What does a Banana Nut Muffin smell like to you? I think it could smell of things like vanilla, banana, nuts, raisins, bread and caramelized sugar.

The more you get to smell things individually, the faster you can pinpoint specific elements to put into your smell memory and flavor vocabulary.

Then the sipping step can prove to be more exciting and more memorable as you test this out with new wines along your tasting journey.

: Savor

Ahhh! The bulk of the work has been done. Now you can sit back, relax and let it all soak in. That's what savor, the last "S," is all about. It focuses on taking in the overall experience as you slowly start to form more concrete opinions about the wine.

The taster should ask themselves the following questions. What does it do to my mouth? What does it taste like after I swallow it? What are the main flavor profiles? What flavors are secondary?

This is also the time we want to contemplate the finish of the wine. You might hear people mention the finish of a wine. The finish is simply just the lingering aftertaste of the wine after you've swallowed.

The finish is an important part of the evaluation. It can give you an idea of the style of wine and/or the quality of a wine. The finish can be described as spicy, bitter, chalky, sweet, smooth or Earthy, just to name a few.

Or it can also be described by time: short, long or non-existent. A good finish can be said to last between 10 to 60 seconds, on average. You would want a wine to have some type of finish that lingers a bit. The duration of how long can depend on the style of wine you have. But you really don't want a finish to be almost non-existent. That could be a sign that the quality of the wine is not great – either from the production of the wine, the storage of the wine or the age of the wine.

Now think about if that wine will be something you would want to enjoy more of or not – and why! Think about the "why?"

Taking this systematic and guided approach to wine tasting provides you a major opportunity to analyze the whole wine tasting experience.

I personally opt for the Five "S" approach because I really like to end the experience at savor. To me, savor is the point where you and the wine really bond.

In this day and age, we get into our daily routines where we're running, running, running. We're multitasking like nobody's business. And we are "phoning in" many of our experiences. We just are not fully paying attention to what's happening in the moment.

We are not present. We are not taking the time, the few minutes, to experience these little luxuries in our lives like food, wine, friends and family.

It's just really nice to push the pause button from time to time. The savor step forces you to do that, while also letting you ascertain what exactly is going on in your glass and on your palate.

: the additional "S's"

While I am a big proponent of the Five "S's," there are a couple additional tactics that can really come in handy when analyzing wine: Spit and Sound.

: spit

This concept is a common practice in the wine world but can come off as a little controversial (i.e. gross) for newcomers to wine. Spit is used by many wine educators and wine professionals as a safe guard against overindulgence and becoming intoxicated. This usually occurs after swishing the wine around in your mouth, then opting to spit the wine out rather than consume it.

When dealing with wine on a regular basis, the practice of spitting is crucial to work success. This is especially true if you attend a large tasting event when there are hundreds of wines. The practice of spitting can let you cast a wider net on what you can successfully sample.

It is also a good practice for people who have a low tolerance for alcohol; the ones who are more susceptible to becoming intoxicated quickly.

Yet when done in public settings it can come off as visually unsettling. The best approach when spitting is to bring the spit cup, spittoon or bucket as closely to your face as possible to avoid others witnessing the activity.

: sound

This isn't a topic many people think about, but it can be a very important part of the learning process.

With wine tasting being such a sensual practice, we want to make sure that we can completely focus on each step of the process. Ideally, the tasting is ideally conducted in a place that is quiet or with a moderate noise level.

We already know that wine can be fun and social and highly enjoyable. Therefore, adding some music to the experience doesn't seem like something that is out of left field. But selecting the wrong music or the wrong volume can be very distracting during the tasting process.

Some companies and sommeliers even have found great success pairing certain wines with poetry, piano, specific songs or a DJ to maximize enjoyment. And like your wine choice, the appropriate noise level in which you enjoy the wine depends on the situation.

However, if you are really trying to focus on the wine in front of you, too much background noise, talking or loud music can be very distracting to the process. That can take your head completely out of the game if you are studying wine.

Just consider the scenario and determine what would work best for you in the given situations.

: the final analysis

This tasting exercise should be fun, enjoyable and empowering. It should not make you sweat bullets. While it might initially fray your nerves a bit, ultimately, it's not a test to see who's the better taster or who has the best senses.

Tasting is very personal. We are all individual. We have different palates. We come from different cultures. We have different backgrounds.

What you smell is what you smell. What you taste is what you taste.

A wine expert can process that and connect you with the typicity of the grape, wine or region. But they can't tell you what you *are* personally experiencing in that moment.

Hopefully the Five "S" process will help open you up to a new world of wine enjoyment and appreciation. Just be open to the experience and enjoy the moment.

CHAPTER SIX: **the personal palate**

If this book is my love letter to wine, then this chapter is a letter of love to my fellow students of wine.

That includes everyone reading this book. Yes, you. YOU! You are officially a student of wine; in case you did not consider yourself one before.

Say it out loud: I AM A STUDENT OF WINE!

Feels good, right! Now that you are owning that, there is something else I want you to own: your personal palate.

Just as the words were appearing on my computer screen – owning your personal palate – I couldn't help but think of OWN as in Oprah Winfrey: the Queen of the OWN empire.

I feel like this chapter gives me my own little Oprah moment where I can write to you: "And this is 'What I Know for Sure!' "

If this was in audiobook form, I would sprinkle some Oprah cadence on it.

So, this is "What I Know for Sure!"

I know wine is a gift.

I know that wine is special and beloved.

I know wine is very personal and individual.

I know that we are all amazingly and beautifully different wine connoisseurs. We come from different communities. We have different cultures. We have different tastes. We have different interests. We are distinctive. We are biochemically different.

I know wine is subjective.

I know perception is a very individual thing.

Everyone reading this book will instinctively have a different reaction to every wine that enters into your lives. You might like it. You might not like it. You might be on the fence about it. That reaction is based on his, her or your personal palate. Own it!

It's like the same way everyone reading this book could have different takeaways from the information provided. We process things differently and we cling on to items – in this case – information that is more relevant to our own lives.

Again, in my Oprah voice: "This is my hope for you."

I want you to get to know your personal palate. You've had it since you were in the womb feeding off everything your mother put in her system. It's yours and only yours.

Learn what it likes in terms of flavors. Learn what it responds to in terms of textures. Learn what excites it. Learn what upsets it. Learn how to make decisions that are pleasing to it.

Listen to it! Honor it!

If you've tried a slew of wine styles from varied regions around the world and all you love are sweet wines. Own it!

If you adore that extremely dehydrating quality that heavy tannic wines offer. Own it!

If you only want to drink red wine with everything you eat from cobb salad to fried catfish. Own it!

Own what you like!

As sommeliers and wine professionals, we've done the leg work to learn the world of wine and how it can be best enjoyed.

By all means, take in what we have to say, write or broadcast about wine. But, even with that, you have to take that information with a grain of salt.

There will be times when our advice will open you up to a whole new world of wine exploration that you never knew existed. You will love and cherish and utilize that advice.

Then there will be times when you don't agree with our advice. Hell, I don't always agree with the advice of some my colleagues or certain wine experts.

That's okay. It's okay to agree to disagree. We are all different. Own it!

Don't allow others to dictate your experience and the experience that's transpiring between the wine and your palate. It's your experience. Own it.

A wine could have a great heritage and impeccable craftsmanship. It could be worth hundreds or thousands of dollars. It could be one of the rarest wines on the planet. It could be *the* most popular wine in the world.

With all that, it's not the right wine for you if you don't personally enjoy it.

You can definitely develop an appreciation for it. I'm all about developing a respect and appreciation for all wines and wine styles. You can appreciate the wine presented to you by someone, but ultimately you want to focus your attention toward wines and styles that are better suited for you.

Wine is just like anything else in life: theatre, fashion, film, music, art, electronics, literature, cars, food.

It's subjective. It's personal. It's individual. An everyone's appreciation levels and interests are different.

My hope is that this empowers you to navigate the next chapter of your wine journey in a way that best suits your needs. The next chapter of this book was tailored to help you learn how to incorporate wine into your life in ideal ways, but on your own terms.

Cheers to owning what intrinsically belongs to you: your palate, your choices, your life.

CHAPTER SEVEN: **the shopping, serving and storing**

Let's be real!

This just might be the chapter that interests everyone the absolute most.

You skimmed the Table of Contents and flipped right to "the shopping, serving, storing" chapter.

Am I right? Or am I right?

No harm. No foul.

I wrote an entire book about wine education from my perspective and from years and years of studying wine and teaching wine classes, but this is what you care most about.

I hope you're laughing along with me. I'm really just being facetious. Or as the French say, "un peu stupide" (a little silly).

I'm flattered you are reading my book in the first place.

While I definitely want you to explore the chapters you are most interested in first. Please revisit the other chapters in the book at your leisure for more context.

Now to the matter at hand: the shopping, serving and storing.

This is probably the most practical chapter in the book. It relates to how you set yourself up to enjoy wine in real life, everyday situations.

I'm a big fan of education, as you can see with my many "soap box" moments throughout the book. Yet many people get frustrated or tune out when the subject doesn't seem to have any real-life application to their lives.

Well, my friends, this is as real as it gets.

The bulk of the content in this chapter was inspired from a collection of questions I've gotten from my wine class students over the years. Many times, a 90-minute or two-hour wine class doesn't allow the fostering of these detailed discussions, so I wanted to put select, key topics into this chapter of the book.

I will make direct connections to how wine can be approached in your daily life while helping you save time and money, and hopefully allowing you to keep your sanity in the process.

: the shopping

Before we go through my top "ins-and-outs" tips of shopping, let's focus on what's really on your mind.

The number one question of all time I get when it comes to wine is PRICE.

How is wine priced? What price should I stay within? If I go up in price will that guarantee me a better wine? Can I get a good bottle of wine under $10?

These are all great and valid questions. So, let's jump right into it.

Price Question 1: How is wine priced?

Pricing is one of the most difficult and complex questions to answer. And, unfortunately, oftentimes the person who asked the question leaves feeling like he or she did not get the answer they really wanted.

If you ask most winemakers about pricing their wine, they would most likely shy away from that topic.

There are so many variables that go into the pricing of wine. It's a bit of a paradox. Just like the wine world, the pricing of wine is not so cut and dry or black and white.

Let's go over four top variables for pricing wine.

Pricing Variable 1 — Economics / The Principle of Supply and Demand

In economics there is a concept called "supply and demand." The supply is how much of a product is available for consumption or sale. And the demand is how "thirsty" consumers are for the products. Sorry for the pun. I really couldn't resist that one.

As we learned earlier in the book, the wine world is just basic farming. These grapes grow on vines on plots of lands throughout the world's wine regions. There is only so much land on which grapes can grow. Wine – the traditional kind – is not manufactured in a laboratory or factory. It's grown from the Earth year after year during each growing cycle.

Due to the size of their operation or the style of wine a winemaker wants to achieve, some producers make very small quantities of wine. These grapes either come from their small plots from each harvest year or they purchase a small quantity of grapes to product their wine in a winery.

Other producers have much larger operations, so a massive amount of wine is produced from the size of their plots of land and / or the purchasing of grapes.

If a small producer has only about 600 cases of wine to sell per year, the demand of that wine could be high since it is so little to go around. Therefore, the price will be set toward the higher end because there is a strong demand for that wine. There is less of it available for general consumption. The more exclusive and rarer the wine is, the more valuable it becomes.

Consumers often times romanticize the wine business. They sometimes can't get past the word *wine*. They forget that this is, in fact, a *business*. The pricing of wine needs to cover business costs and ideally make a profit at some point.

Then there are times, especially in the New World wine regions, where we have mass producers of wine. A producer of this magnitude could produce say 600,000 cases of wine each year. Because the supply is so high – even with a high demand – the prices would be on the lower spectrum. The more accessible the wine, the less inexpensive it becomes.

These production styles also factor into the "taste of the wine" and the consumer's perception of wine. Think about the varied nuances of wine from the place, to the grapes and to the production styles.

You are apt to get a more distinctive wine garnered from smaller producers. There is a lot more care, oversight and focus on the grapes and the land in which they grow. There could also be more of an artisanal approach to the winemaking with small producers. These wines will therefore give you a stronger sense of terrior – that sense of place in the bottle.

As fine and lovely and delicious as a bottle – or a box – of wine from larger producers, these options will never be as complex as those of the small production wines. There are so many grapes in which to account for and look after over acres and acres of land.

And with the grapes coming from several varied plots of land in a larger area, you lose a lot of that "sense of place" character in the wine.

Therefore, a major factor in pricing is supply and demand.

Pricing Variable 2 — Cost of International Business

The second most-asked question about price is why wines in the United States are so much more expensive than when you buy them overseas.

For pricing of wine – for me – all roads lead back to "supply and demand." But also considering that wine is a global business, there are additional factors to consider that attribute to the cost of doing business around the world.

Let's go back to that concept of an artisanal style that's more associated with smaller producers. If you visit a village in Tuscany, Italy, or take a biking wine tour in Loire Valley, France, or spend the summer drinking wine near the beaches of Barcelona, Spain, you are most likely getting wine that is produced very close to where you are enjoying the wine.

If a wine is produced by a small, local winemaker, he or she typically sells the small amounts of wine to nearby regions at a very reasonable price. There is little or no need to do advertising and promotion to sell the wine. And the wine production is so small, that it never sees the inside of an airplane or a ship to be transported around the world.

Therefore, the costs of doing business are lower. That helps make the price of wine more affordable and accessible.

On the other hand, let's look at those previously mentioned regions that supply products to their local areas but also export their products around the world. The consumers on the receiving end, in different parts of the world, will pay the higher price for the cost of doing business to have access to the products in their local stores.

Wine can be romantic, but it's still a business.

Winemakers that export their wines to outside areas have to deal with cultivating new markets, hiring licensed importers and / or distribution companies to get and sell their wines in other countries, pay for new packaging, pay the overseas shipping on products and pay the insurance on products. Then they have to take into consideration different exchange rates, inflation rates, taxes. On top of that, they have to pay to sustain sales via marketing events, in-store sampling and promotions.

And that, my friends, are some of the reasons why a wine sold in the United States will be more expensive than what you pay for while you are on vacation in another country.

Pricing Variable 3 — Old World Prestige

Wine carries a lot of prestige. This is especially true when you have highly sought-after wines from some of the most classic wine regions of the world.

When you think classics in the wine world, you think Old World. And when you think Old World classics, France and Italy are the two countries that stand out most.

France and Italy are in this continuous struggle to be the top producing country of wine, in terms of volume, in the world. The two go back and forth from holding the number one position every few years.

However, when it comes to the most expensive, luxurious and sought-after wines in the world, France wins that battle hands-down.

A large portion of the world's most expensive, prestigious and sought-after wines come from Bordeaux and Burgundy, France.

A bottle of Chateau Petrus, a red wine from Bordeaux can cost customers anywhere from $2,600, on average, per bottle. Then you go over to Burgundy and a single bottle of Pinot Noir from Domaine de la Romanée-Conti (DRC), can garner prices between $2,000 and $20,000.

What makes one bottle of wine like either of these examples so valuable?

It's the prestige element.

It's the tradition. It's the cache. It's the land. It's the consistency. It's the level of expertise. It's the art. It's the science. It's the exclusivity. It's the rarity. It's the intrigue. It's the grapes in an ideal terrior.

And, in most cases, it's the protection by the French government to require a high-level of quality from these producers and dictate how much wine can be produced from these areas.

These winemakers are producing a small- to medium-sized amount of wine, so there is always little of it to go around. And that connects back to supply and demand.

Pricing Variable 4 — New-World Cult of Personality

What New World wine regions lack in "classic" terrior and legacy, they more than make up in terms of celebrity winemakers and brands with a cult following.

That's not to say that New World wine regions don't have amazing terrior. Of course, they do. These pieces of land produce some of the world's most brilliant, amazing and striking wines. It just that these regions don't have that classic prestige level garnered from time in the game.

What New World wines lack in tenure and tradition, they make up in swag and savvy. These regions have developed prestige in other ways in the form of celebrity winemakers and cult classics.

It's the renegade nature. It's the embracing of technology. It's the vision. It's the pioneering spirit. It's the intellect. It's the revolutionary behind the wine. It's the innovation. It's the passion. It's the rigor. It's the genius. And it's the freedom to create whatever they want.

A few brands that standout in these terms are Harlan Estate, Penfolds in Australia, Opus One Winery, Screaming Eagle, Joseph Phelps and Sine Qua Non.

Going back to a small supply of wines produced with an extremely high demand, these producers have carved out an incredible niche for themselves that transcends "classic prestige" and propels them into a cult status all their own.

Price Question 2: What price should I stay within?

Drink what you can afford.

If you can afford a $3,000 bottle of Screaming Eagle from Napa Valley, then by all means! Cheers!

However, you shouldn't have to take out a second mortgage to enjoy good wines.

And if you can afford a $50 bottle of wine from time to time, don't always regulate your wine choices to the $10 and under bin either.

Think of wine consumption in terms of moods, circumstances and situations.

What's the occasion? Are you having a bottle of wine on a Tuesday night while decompressing from work? Or are you planning on binge watching every season of the "The Unbreakable Kimmie Schmidt" on Netflix over a three-day weekend?

Maybe those wine choices would be some options around the $10 to $15 range.

Think about the situation. It is casual. It is light. You want to have fun or relax or both. So, you'd want to go for the inexpensive, yet delicious options that are great to drink knowing the wine is not the main attraction.

Quality wines from regions in this price range could be from places like Chile, South Africa, Paso Robles, Germany and Australia.

Now let's consider another scenario. Are you prepping dinner for a third or fourth date? Are the guys coming over to help you map out your strategy on the road to becoming an entrepreneurial? Did you want to sip on something nice just because you can?

Maybe you elevate the options to the $15 to $35 range.

Think about the situation. It's more engaging. Wine is part of the experience more. It's a way to treat yourself. You'd want to up the ante a little as the wine will have a bigger role in the occasion, bringing some lovely attributes to the table.

Quality wines from regions in this price range could be from places like Argentina, Sicily, Sonoma, Loire Valley, The Rhone Valley, Oregon and New York State.

Let's go over one more scenario. Is there something to celebrate? Did your wife get that long overdue promotion at work? Did your son secure that fellowship to help continue his studies? Is Grand Dad/Big Papa having a significant birthday coming up?

Maybe you reach for the bottles in the $35 to $100 range.

Think about the situation. All wines can be used for celebration, but not all wines are celebratory. You'd want to make the special situation standout with a special, memorable wine.

Quality wines from regions, in this price range, could include places like Champagne, Napa, Bordeaux, Barolo, Burgundy, Rioja and Tuscany.

Try to think about the context in which you are shopping for a wine and find suitable options to meet those needs within your budget and preferred wine styles.

Price Question 3: If I go up in price will that guarantee me a better wine?

Putting some extra dollars toward a bottle of wine could bring you closer to a product that has received an elevated level of care, attention, experience and expertise by the winemaking team.

In that regard, we can say that you have a better-quality wine on your hands by upping your price range.

However, higher priced wines don't automatically translate into if you will personally enjoy or like the wine more.

Let's go back to the level of care, attention, expertise and experience of a winemaker. This winemaker is going to produce the best possible product to showcase the land, the grape and either tradition, innovation or both.

Those attributes are going to lead to a more distinctive, layered and multifaceted wine. That sounds amazing to me. But for many people, on the palate, it doesn't translate to delicious.

That's when a little research goes a long way. CHAPTER FIVE outlined ways in which to get better acquainted with the textures and flavors and structures of wine that are better suited for you.

Try to get as much information about the wine before you purchase it to see if the wine's tasting notes and descriptions link up to what you tend to like. Also, see what other shoppers say about the wine. There may be some grey area here. But if you see things that are contradictory to what you enjoy, maybe put those wines on the back burner for now.

Ultimately, to answer this question, price is not a determining factor on whether you will like a wine. Higher priced items are more of a gamble, especially if you are not familiar with the style, aromas and tastes of the wine you are considering.

Simply, drink what you like. Pay for what you can afford. Do your homework before making a splurge purchase.

Price Question 4: Can I get a good bottle of wine under $10?

Abso-freaking-lutely! You can definitely get a lovely bottle of wine for under $10. Just like shopping for slightly more expensive wines, you will have to do some research and a lot of digging.

I love this price category! When I first started working in the wine world, the shop I worked for on The Upper West Side of Manhattan touted an array of wines for "$10 and Under."

The customers loved it. The employees loved it. I loved it.

This is one of the best ways to jump start your learning about wine by tasting. And nothing expands your tasting options like the affordable under $10 options.

Now, I know some of you have raised eyebrows and crinkled foreheads right now. You might be questioning my credentials for the first time or for the 10th time in this book. But, again, let me explain.

I know there might have been a time or several times when you purchased a wine for a Monday night, paid $8.99 plus tax for it, and absolutely hated it.

Those wines are out there. The problem is either it's just not a quality wine or it just doesn't stand up to your personal palate.

BUT! But! But there are some amazing options out there. Just approach it with an open mind and an adventurous spirit.

Take a chance! With the cost of these bottles being so low, you are not at a major loss.

Maybe it's a huge success. You can find your new Monday, Tuesday night wine or binge-watching wine.

Perhaps it is just wrong. The wrong flavor profiles. The wrong acidity. The wrong finish. Just wrong!

That's a win too.

At the end of the day, you would have tried something new. You've given your palate another perspective in which to analyze wines – to better understand what you don't like.

If you don't like it, you can always repurpose the rest of the bottle. Use it as an extra seasoning for your ragù sauce or chili or meat loaf. Marinate meat with it. Or turn it into a fun little Sangria or wine spritzer.

The moral of this answer is do your research. A little due diligence goes a long way. Trust me! I've done the leg work for several years.

Here are a few overall tips when shopping for wines $10 and under:

Tip No. 1: Find a reputable shop that you trust with all your wine purchases

Tip No. 2: Find wine regions that consistently offer low prices and high quality

Tip No. 3: Find a way to have fun with the experience no matter what happens

: the wine shops

When it comes to shopping for wine in general, you should select a few places to frequent that you know, trust and respect.

You can build a trust in their selections by visiting them often for in-store tastings. It's always great to try before you buy – when at all possible.

Then you can test out the wines you are familiar with that they happen to carry. You want to make sure the wines are taken care of and stored properly at every level. Test out their lower priced inventory to determine if its good quality in addition to a good value.

Lastly, you should try to develop a rapport with the staff. Ask questions. Ask for recommendations. Get a gauge on how serious they are about their wine store and the types of products offered to customers.

Remember! Wine is a business.

Some people love wine and find a way to make it work as a business. And some people love business and find a way to make it work through wine. There is a big difference.

: the affordable wine regions

There are definitely regions that have the "Old World" prestige. Then there are the producers that have the "cult" status.

But then there are some amazing wine regions around the world that are known to offer up great value and quality. You want the two – quality and value – when shopping for wines in the lower price ranges.

The key is to shop for some of those "under the radar" regions. They are often found in some classic winemaking countries. Then there are some wine regions around the world that are really trying hard to make a name for themselves, so their wines are priced competitively.

These are some of my go-to regions, many of which offer quality wine for $10 and under:

CHILE

Appearing globally on the wine scene a little later in the game – compared to most wine regions – the Chilean wine world is still growing, developing and coming into its own. That makes Chile a very exciting market to watch. But for consumers, it gives them the great opportunity of obtaining lovely wine at very reasonable price points.

Look for delicious and affordable Chardonnay, Cabernet Sauvignon, Merlot and Sauvignon Blanc. And, from the affordable to the mid-price range, try Chile's signature red wine Carménère. It's a beautiful wine with lush hints of smoked plum notes.

These wines can range anywhere in price from $7.99 to $20.99

SOUTH AFRICA

While the wine industry in South Africa emerged well into the 1600s, the country had been haunted globally by the backlash against its politics in relation to the Apartheid system. The strict racial segregation in the region, starting in the 1940s, was met with fierce international pushback in the 1970s. That put a hold on any wines from South African being exported to countries that disagreed with the policy. Ultimately with Apartheid being repelled in the mid 1990s, that finally opened South African wine to the global market place. As a result, it is considered an "up-and-coming" region. It is still trying to gain its foothold and leave its distinct fingerprints on the wine world. That helps bring forth lots of values in terms of pricing.

Look for the amazing red blends, white blends, sparkling wines, Merlot, Shiraz, Cabernet Sauvignon, Chardonnay and Sauvignon Blanc. One of the stand out options is the Chenin Blanc grape, also known as Steen in South Africa. It's adored for its vivacious expression of citrus, melon and stone fruit. And then South Africa's signature red wine, Pinotage, is a highly distinctive option from the inexpensive to the high-end.

These wines can range between $9.99 and $20.99

SPAIN

As one of the classic wine regions in Old World Europe, Spain straddles the fence in terms of production styles, innovations and winemaking philosophies. Producing more than five billion bottles per year on average, many of those wines come from outside the standard, esteemed regions of Rioja and Priorat. Wines from other regions like La Mancha and Calatayud are more plentiful in terms of production and sometimes sold as "bulk wines." That keeps the price low while still showcasing Spain's high-quality heritage.

Look for great Tempranillo, Garnacha, Graciano, Monastrell, Airen, Viura, Verdejo and sparkling wines.

These wines can range between $8.99 and $20.99

SOUTH OF FRANCE

France is synonymous with luxury wine. From Champagne and Burgundy to Bordeaux and The Rhone Valley, these regions represent some of the most high-end wines on Earth. However, in a large stretch in the South of France – Sud de France – there lies a selection of big, bold, budget-conscious wines, from reds to whites to sparklings. First planted by the Greeks and then propelled forward by the Romans, this is an area that encompasses the Languedoc-Roussillon region and houses more than 300,000 hectares of land. That's a lot of wine being produced and, therefore, a lot of value pricing.

Look for Syrah / Grenache-based blends, Merlot, Bourboulenc, Chardonnay and Chenin Blanc-based sparkling wines.

These wines can range between $9.99 and $25.99

: the shopping — the label

Stepping into a wine shop – large or small – can send you into sensory overload. There are dozens, if not hundreds, of options to choose from. Store shelves can be separated out by various categories: region, varietal or style (crisp, rich, mellow, bold).

As you peruse the selections, you might be attracted to the familiar or the pretty label art. You go in for closer inspection to get more information on the label to see if it'll be an ideal fit.

Labels have a big job to do. They are trying to incite a sale based on a number of strategic factors. Labels are also there to provide some much-needed information, some more than others. Some information is mandated by law. These are items you will find on a typical label:

Country of origin — Since the land is very important in the winemaking process, the label will state the country in which the wine was produced. That gives you an idea of the style of wine known from that region. The countries could be from very classic wine regions, like Spain and Germany, to little known wine regions India and Croatia. Along with the country of origin, the region and sometimes sub regions are also stated on the labels.

Varietal — The varietal of the wine will oftentimes be stated on the label of the wine, most commonly with New World wines and select Old World wine regions. Most Old-World wine regions like Italy, France and Spain, will not feature the grape varietal on the label and only state the country of origin and the region the wine comes from like Barolo, Bordeaux and Ribera del Duero. In this regard, the land is the most important aspect in the winemaking process. To know what grape varietal or varietals you are drinking in the bottle, you'll have to know what grapes grow in those regions. With today's technology, an APP or a quick online search of the wine or the wine region will give you that information.

Vintage and Non-Vintage — The vintage is the particular year the grapes were turned into wine. Riesling grown and produced in 2000 is therefore a 2000 vintage Riesling. This next part is where some confusion sets in. Most still wines, non-sparkling, will have a clear vintage stated on the label. There are some countries, like Portugal, who will blend multiple vintages together, bottle it and sell it. Since there is no one clear vintage in which the wine comes from, the wine is referred to as a Non-Vintage wine. The same practice is common for most sparkling wines, like Cava,

Champagne and Prosecco. The winemakers blend multiple vintages of wines together to create their signature style of sparkling. In select cases, sparkling wine will have a vintage if the winemaker deems a particular year as ideal for having growing grapes. If that happens, only grapes from that one particular year are used to make that sparkling wine. That sparkling wine is therefore a vintage sparkling. Vintage sparkling wines are rarer, highly regarded and will be more expensive than their non-vintage counterparts.

Brand name and / or name of wine — Brands are an important part of the label. It distinguishes who is actually producing the wine. Some brands make multiple wines from the same grape varietal. When that happens, the wine is referred to by the brand name, the grape varietal and vineyard plot name and/or a special name for that particular wine. For example, Stag's Leap Wine Cellars is a brand in Napa Valley, California. The brand makes a variety of wines from the Cabernet Sauvignon grape. However, the brand has different names of their wines made with the same grape to distinguish between the options. For example, there is Stag's Leap Wine Cellars *Artemis* Cabernet Sauvignon, Stag's Leap Wine Cellars *Fay Vineyard* Cabernet Sauvignon and Stag's Leap Wine Cellars *Cask 23* Cabernet Sauvignon. Therefore, the label features the brand – Stag's Leap Wine Cellars – and the different names associated with the wines produced under that one brand.

Importer and / or distributor —The importer and distributor are the companies responsible for bringing the wine to the market through certain distribution channels. Importers source wine from wineries around the world in the hopes of having the wine sold in different markets. Distributors are companies that work – through the use of sales reps – to place wines in retail stores and restaurants. There are some importers and distributors who have gained a high level of respect in the industry like Frederick Wildman & Sons or Jenny and Francois Selections. Therefore, some consumers will check the label for the importer and / or distributor they trust to finalize their purchase decision.

Alcohol by volume (ABV) — The alcohol content of the wine, also known as Alcohol By Volume (ABV), is always stated on the label. It is a mandatory government requirement for that information to be stated for the consumer's review to aid in responsible drinking. It is also because alcoholic beverages are taxed according their ABV. Average alcohol contents range between 12 to 14 percent. That number can definitely be lower with wines as low as 5 percent alcohol and much higher with table wines reaching 16 or 17 percent in alcohol.

Net contents — This information is provided to showcase just how much wine is contained in the bottle. The standard wine bottles are 750 ML (milliliters) in size. That holds three-quarters of a liter of wine (about 25 ounces or five, 5-ounce glasses of wine).

Government warnings — With alcohol being regulated by governmental agencies around the world, consumers will always find a mandatory warning label on each wine purchased. These warnings are meant to safeguard consumers from harm.

Contains sulfites — Based on government regulations, any item containing 10 ppm (parts per million) and up to 350 ppm of sulfites in the U.S., must be stated on the label. Most wines average 80-120 ppm.

There are a few things, however, that you won't find on the label: expiration dates, calorie counts and carbs.

Expiration Date — As stated previously, wine is a living, breathing entity. Each wine bottle is its own entity and will age at its own rate, especially given the storage conditions. Therefore, there are too many variables in which to state a "use-by" date on the label. The vintage information on the bottle – coupled with knowing how the grape varietal ages – can allow you to gauge the "best time to drink." Wine publications and vintage charts are available to determine "when to drink" for select wines.

Calorie Count — Wine has a different governing body than other items that fall under the purview of the FDA (Food and Drug

Administration) in the United States. Wine falls under the jurisdiction of the Department of the Treasury's Alcohol and Tobacco Tax and Trade Bureau in the United States. Therefore, wines don't require a nutrition label like milk or juice or packaged food stuffs. The only time you will see a nutritional label on a wine bottle is when the bottle is a "wine-product." That is a wine-based beverage in which other items have been added to the wine like sugar, juice and flavorings. However, a typical serving (5 ounces) of a dry table wine comes in between 100 to 150 calories. That would be about 500 calories to 750 calories per bottle.

Carbohydrates — The same principle applies to carbs as it does to calories in wine. It's not mandated to be put on a label unless the bottle is a "wine product." However, there are typically 4 grams of carbohydrates (carbs) in a 5-ounce glass of dry red or white table wine.

: the shopping — the vintage

The vintage of a wine can be very perplexing for shoppers. Essentially, the vintage of a wine is the year in which the grapes were turned into wine. Grapes have a growing cycle every year, in both the Northern and Southern Hemispheres. The year the grapes are picked and turned into wine is the vintage year. If grapes were picked in 2011, then the wine is a 2011 vintage. If grapes were picked in 1966, then that is a 1966 vintage.

The confusion oftentimes sets in when *one* word has *two* different meanings that are applied to the wine world. In this case, it is the word "vintage."

Sometimes wine is referred to as a "vintage" wine. In that context, that means it is an older wine. A 1966 Cabernet Sauvignon from Napa can be "vintage," stating its status as a much older wine also while indicating that is comes from the 1966 vintage.

But a 2016 Cabernet Sauvignon from Napa – from the 2016 harvest – cannot be referred to as "vintage" in terms of age because it is not very old. But it does represent a wine from the 2016 vintage.

Therefore, in the wine world, vintage most commonly refers to when the grapes were turned into wine during the yearly harvest. Vintage in terms of retro or older or antique is best reserved to describe cars, clothing and furniture.

Many consumers have a strong preoccupation with the vintage of wine when shopping. It is my belief that it goes back to prestige and the old adage that "wine is better with age"

While that is definitely true of some wines, the "better with age" notion does not apply to all wines. And therefore, a customer's fixation with vintage does not always align with their specific shopping needs.

I know. I know. What does that mean?

Please don't get me wrong. The vintage of the wine is *absolutely* important for so many reasons. But as we've learned previously, not all wine is meant to age for a long time or is considered to be "age worthy."

Understanding this notion will save you a lot of time and money when shopping for wine.

I really want people to understand that wine is a living, breathing product. While wine doesn't come with an "expiration date" on the label like many other products you can purchase, all wine will expire in the bottle at some point.

Some wines will expire sooner than others. Take Pinot Grigio, Sauvignon Blanc and most rosé wines for instance. These are some wines that are meant to be consumed quite young – one year to about three years of the wine's vintage in some cases. The potential of these wines will peak early, and the wine starts to decline in character and structure in the bottle.

Then there are some wines that are "age worthy." These wines need a generous amount of time to mature properly or "come into its own" so to speak.

These wines offer more concentration of flavor. They are denser. There is some power there. Over time these wines will soften up a bit and become more balanced. In terms of red wines, these styles are considered too harsh to drink when they are young because of their tannic structure. They will take some time to reach their full potential. During that time, the firm tannic structure – the quality that dries your mouth out – changes to a softer, more approachable structure.

Knowing the vintage can give you better insight into if you should buy the wine at all, buy it and drink it right away. It also lets you know if you should buy it and let it rest for some additional time.

: the shopping — the older vintages

Let's revisit that 1966 Napa Valley Cabernet Sauvignon that was mentioned earlier.

That sounds amazing, right!

I totally understand the fascination with tasting older wines. Older wines represent prestige. Cellar collections. Wealth. Being wealth-adjacent. Rarity. Status.

There is a certain cache there.

Let's imagine that 1966 was a stellar year for grape growing in Napa Valley. Let's choose the Oakville region of Napa to be specific. All the stars aligned. The grapes received the right amount of sun. The right amount of rain. The right amount heat during the day time and the optimal cooling temperature at night.

Sampling wines from decades past can indeed be a thrilling experience. It's like opening up a time capsule. You are getting an actual taste of that specific time, production style and place in the bottle.

It's full of surprises – good or bad. Surprise!

If we were to open that 1966 wine right now, it would be more than 50 years old. A Cabernet Sauvignon from that region would definitely have the needed structures – acidity, tannins and alcohol – to age well over time.

But a wine with this much age is now fragile. It still has some power, but it's not as strong as it used to be. Some of the structures will not hold up as much as they used to back in the day. That's just the reality of life. Things fall apart.

Since wines are living things that are constantly changing, no two bottles of wines are going to develop at the same rate. It doesn't matter if it's from the same plot of land or the same year or from the same barrel. Each bottle is its own entity.

But take the risk. Take the gamble. Every purchase of wine is a gamble if you think about it.

However, please don't be surprised if you do not like it. Older wines can be an acquired taste. On the other hand, you just might hit the jackpot. You'll never know until you open it and try it.

When in the market for older wines or wines to lay down to rest, it is important to consult a vintage chart. These charts can really come in handy with moderate to higher-end wine purchases. Vintage charts are often produced by magazines or some organization by analyzing the wine, the weather conditions of the harvest and intel from the wine growers from that region. You can also find some online.

Consulting these charts will give buyers a timeframe in which to let the wine rest and a ball-park window on when the wine should peak and be opened. Otherwise, if the wine is past its prime when you purchase it, you might have some very expensive vinegar on your hands.

: the shopping — good and bad years

The wine growing cycle experiences a series of ups and downs every year as outlined in CHAPTER TWO. There are so many factors and variables on what can be deemed good or bad including erratic weather conditions like drought, hail storms, frost, floods, earthquakes, wild fires.

Even with all that, a great winemaker is capable of making a good wine despite not having the ideal growing conditions. The downside of that fact is there will just be less good wine from that region produced that year.

For further study on this matter – particularly when making a deeper investment on a wine purchase – vintage charts are a great resource.

: the shopping — the giving of wine gifts

"It is not the gift, but the thought that counts." That's a famous quote by American author, poet, diplomat and theologian Henry van Dyke.

This is a concept that many people use to encourage a person to be thankful for the gift even if they don't personally like it.

In some regards, I do subscribe to that philosophy. Being thoughtful is wonderful. Thoughtful people are my favorite kind of people. And a person should appreciate what has been given to them.

I have one major question regarding that train of thought, however. Exactly how much thought was put into the gift?

In my opinion, it is not just enough to just think about giving the gift. The gift should speak to the person's interest and preferences. When it comes to wine, this is especially true, because wine is very personal to people.

Throughout my career, I've been asked tons of times to help pick out a wine for a birthday present or Christmas present or as a business "thank you" gift.

When asked what type of wine the recipient might enjoy, the responses range from "he likes red" to "can you just suggest a $75-$100 bottle of wine for an older couple who loves wine?"

I'm not trying to be difficult here. But with wine being as personal as it is, I will definitely need a little more information than that.

Don't get me wrong. The heart is in the right place. And I love the fact that they want to give the gift of wine. My suggestion, though, is just to do a little more research to see what wine would be most appropriate as a gift.

Christmas and Birthdays come around once a year. Try to do a little digging or subtle questioning about their preferences and try to match that as closely as possible for your wine loving friends and family and colleagues.

You wouldn't want to give a bottle of California Cabernet Sauvignon if the recipient only drinks French Pinot Noir. While they are both lovely red wines, one might not be as appealing as the other based on the person's wine preferences.

Again, the thought is there. Absolutely. Unfortunately, the wine may be of no use to the recipient and the gift will be wasted on them if he or she drinks the exact opposite of what was gifted to them.

Don't shoot the messenger. People like what they like. I'm only trying to help all parties involved.

If you are not sure what would make an ideal gift, gift cards to their favorite wine store or a well-respected store make for a very thoughtful and stress-free gift. If you pair that with a nice note or card, they can pick what they like, and they'll have their wine and can enjoy drinking it too.

: the shopping — the milestone gifts

The other common mistake people make when selecting gifts is trying to find wine from a year that marks a major milestone.

I often have people – typically family members – looking for a wine from the recipient's birth year for a birthday or wedding gift.

My advice is to think twice. And then maybe think it over a few more times. Then a few more times just to be *completely* certain that's what you really want to do.

I know I might be coming off as a real Debbie Downer in this section about gifts. Honestly, I'm really trying to save you the time, effort, expense and potentially heartache of making a risky gift choice.

The first challenge is trying to find a wine on the market that is 20-plus years old. Remember there is only so much wine produced from each harvest every year. Once that wine is gone from that vintage year, there is no way of ever getting it again. It may no longer be available on the market.

If you are able to find some older vintage wines that match the birth year, the second challenge is getting over the sticker shock in terms of price. The older the wine, the rarer it is going to be. Therefore, the wine will cost you a nice chunk of change.

On top of that, older wines can be an acquired taste. As wines age, they will lose a lot of the generous fruit notes of their youth and tastes more of dried leaves, wet wood, gravel, dried fruit and Earth, in many cases.

At the end of the day, a gift is meant to bring someone joy. This is not a time to take a major risk. Buying wine is risky enough. This is one of those times when it is perfectly fine to play it safe with a more calculated risk.

: the shopping — corks or screw caps?

When buying wines, the type of closure a wine bottle has can be an immediate deal breaker for some consumers.

I understand the apprehension. I get it. Cork is familiar and comforting to wine lovers. There is a certain elegance, a certain cachèt, a certain "*je ne sais quoi*" (*I don't know what*) quality associated with opening a bottle of wine with a cork stopper. It is also the "Pomp and Circumstance" when it comes to the wine bottle presentation at a lovely restaurant or during a special occasion.

Seeing a screw cap doesn't visually connect to a "quality wine" for some consumers. For them, it screams "cheap" or "unsophisticated" or even "gauche."

Let us delve deeper into what these two types of bottle closures are and how these facts might factor into your decision-making process.

The overarching goal of cork and screw caps is to seal the wine away from oxygen and other elements: dust, dirt, debris.

Cork is a natural material shaved off of the bark of two different types of oak trees that grow in regions of the Northwest part of Africa and in Southwest Europe. The lion's share of cork produced comes from Portugal, about 300,000 tons worth.

It is something that has been used in the wine world since ancient times. Historians have found it in tombs dating as far back to ancient Egypt as stoppers in clay bottles. Its use become even more common during the Dark Ages in Europe when wine was safer to drink than water. And, of course, it is the main style of bottle stopper that comes to mind when we think about wine.

Besides it being a mainstay in the wine world, the properties of cork are beneficial to the long-term storing and aging of wine. Since cork is porous – full of tiny spaces – it allows for micro-oxygenation or the flow of oxygen into the wine.

That helps the wine mature and develop over time. With wines that can actually benefit from extended aging like Brunello di Montalcino, Chateauneuf-du-Pape, Barolo and Bordeaux, cork will probably always have a role in helping those wines progress over time.

However, with cork being a natural material, there are some downsides that come with using cork to seal off wine bottles.

The first major issue is the risk of TCA taint or "cork taint." This was discussed in more detail in CHAPTER THREE. This essentially occurs when the TCA chemical compound leaches unfavorable smells and flavors into the wine that might be associated with wet newspaper, damp cloth or a mildew in basements.

Then there comes the fragility of cork when improperly stored or from extended aging. Cork can easily dry out if the bottle is stored improperly over a long period of time. Once that happens, the cork can fall apart when being removed by a wine key / corkscrew. Then you have to think of clever ways of getting the cork out, either using a two-prong cork puller or push the cork into the bottle. While the latter is not harmful for the wine, it does make fishing those pieces out of the wine or glass annoying.

Lastly, cork has to be shaved from trees. If overdone, it can have negative environmental effects and result in conservation and / or reforestation efforts.

Next up we have the synthetic corks. These would be the middle ground between traditional corks and screw caps. It's an alternative that offers up some of that ceremonial pleasure while guarding against the potential contamination of the wine by TCA.

Then there is the new kid on the block: screw caps.

These Stelvin metal screw caps feature a perforated top and a long outside skirt. This skirt resembles the foil on the bottle of a cork stopper. The goal of these screw caps is to make opening easier, protect against contamination, preserve aromas and freshness, and ensure a quality product delivered to consumers.

However, these types of closures were not very well received initially. When the wine world, a few decades back, was undergoing a renaissance of sorts globally, the focus was on quality, prestige and luxury. It was a departure from the casual consumption of wine in the 1960s and early 1970s.

After being reintroduced in the late 1980s and early 1990s, the concept of screw caps on wine started to stick a bit. New World markets like Australia and New Zealand were the early adapters of using screw caps as the standard closure for the majority of their wines. Usage of screw caps nearly doubled and tripled in some cases in those areas.

Some of the classic and well-respected winemakers in France and Italy are now fans of screw caps for their wine. Domaine Laroche, a wine producer in Chablis, France, has deep roots in the Old-World winemaking culture of more than 1,000 years. Yet is also has its eyes on the future. In 2001, the producer started bottling its Chablis, Premier Cru and Grand Cru with screw caps. Producers in

Sancerre, France have followed suit to keep their Sauvignon Blanc offerings fresh and lively. And even some red wine producers in Veneto, Italy, have bucked the system and opted to use stew caps.

As a person of a certain age, you eventually get more accustomed with the notion of change. Maybe not totally comfortable, but you get more and more used to it. I've seen the shift of enjoying music by vinyl to 8-track tapes to cassettes to compact disks to digital files during my lifetime. Enjoying music on the go shifted from the Boombox to the Walkman to the Discman to the MP3 player to the iPod to the mobile telephone.

Things change. As things in life change and new technology arises, it is best to keep an open mind, or you might get left behind.

I don't envision that screw caps will ever totally replace cork stoppers made from trees. I believe cork stoppers will always be around, especially in connection with those wines that benefit from resting and aging and maturing in the bottle for years and years.

The options are there, however. It's a beautiful thing to have options. Despite the type of closure – cork, synthetic or screw cap – it's what in the bottle that's most important.

And it is up to you, personally, to determine which option is best for you and your particular occasion.

: the shopping — bottle sizes

Knowing the size of the wines, especially in relation to your shopping needs, can be very helpful.

The most common wine bottle is the 750ML size. And yet wine bottles range from the single-serve size to very large formats to suit certain occasions.

A larger bottle of wine is great for special occasions and parties. It can definitely equate to more fun and the opportunity to serve a larger number of people. There's no question about that.

However, the size of the bottle will also equate to the wine's individual preservation. Oxygen is a friend and enemy of wine. A sufficient amount of oxygen helps the wine develop in the bottle over time, while too much oxygen makes the wine spoil.

Wines will get small infusions of oxygen through the cork and through some Stelvin screw caps, given the oxygen transfer rates built into the metal cap. The smaller the ratio of wine to oxygen, the faster the wine will spoil.

When it comes to the long-term storage of wines, Magnum bottles (1.5 Liter size) tend to age better or slower than the standard size bottle. There is a larger volume of wine in the bigger bottle, but the cork size is the same as the 750ML bottle. As a result, the larger bottles get less oxygen and will age much slower than the smaller bottles.

Here is a quick look at the sizing options on the market, many of which are named after kings and people found in the Bible:

- **187ML** — This size is known as the Split. It is a single serving size of wine in still and sparkling options.

- **375ML** — This is a half-bottle size that offers up 2.5 glasses of wine. You can find just about any style of wine in this size – non-sparkling and sparkling.

- **750ML** — This is the standard size readily found on the market. It offers about 25 ounces of wine that measures out to about 5, 5-ounce glasses of wine.

- **1 Liter** — This size of wine isn't very common but can be found from select European and New World producers. This size offers up about 7 glasses of wine, one-third more than the standard size.

- **Magnum** — The most common of the larger formats, this size features 1.5 Liters of wine. That equals two standard bottles and equates to about 10 glasses of wine.

- **Jeroboam / Double Magnum** — Named after a Biblical king, this is twice the size of the magnum at 3 Liters. That equates to four standard bottles and up to 20 glasses of wine.

- **Rehoboam** — This bottle comes in at 4.5 Liters. That holds six standard bottles and offers up to 30 glasses of wine. These are mostly reserved for sparkling wine options. The numerical size of this wine causes some confusion because this size of wine in Bordeaux, France, is called Jeroboam. Just be sure to specify what type of wine you are looking for in this size.

- **Methuselah** — Named after the oldest man in the Bible, this is a 6-Liter bottle of wine. That equates to 8 standard bottles and up to 40 glasses of wine. This 6-Liter size also causes confusion when shopping as it is called an "Imperial" in Bordeaux, France. Again, just be sure to specify what type of wine you are looking for in this size.

Surprisingly, that is not where the larger formats stop. These bottles can continue to go as big as 20 Liters in size. When shopping for these larger formats, it is important to point out that many of the larger sizes can only be secured through a special order with a winery.

: the shopping — bottle shapes

The shape of a bottle is one of those elements that many people don't necessarily focus on or, for that matter, even notice. But there are some fun facts about the bottle shapes that reflect back to where the wine is made and, in some cases, speak to the functionality of the bottle.

Here is a brief overview of the major wine bottles shapes:

Alsace Flute — These bottles are tall, thin and represent wines that come from the Alsace region of France and from Germany. That region has gone from both French and German rule over time and has been a part of France since after World War II. The bottles are called either Flûtes d'Alsace or vin du Rein (Rhine wine bottle), referring to both regions respectively. This bottle shape can also be found from some Austrian wines. Wines associated with this bottle shape include Riesling, Gewurztraminer, Gruner Veltliner, Pinot Blanc, Pinot Gris and Pinot Noir.

Champagne — All wines with bubbles are contained in a special sparkling wine bottle. These bottles hold Champagne, Cava, Prosecco, Sekt and other sparkling wines from around the world. This style of bottle is thicker and heavier than still, non-sparkling wine bottles. That is because it needs to successfully contain the high-pressure contents of the sparkling juice. That pressure can range from 60 to 90 pounds per square inch (psi). That is the equivalent to the pressure found in the tire of a double decker bus or two car tires. English glass production used coal-fueled ovens to produce more durable glass bottles during the 17th century, which gave birth to this type of bottle.

Bordeaux — The shape of these Bordeaux bottles is slender with noticeable shoulders near the neck of the bottle. They were used to contain wines made from traditional red Bordeaux blends featuring Cabernet Sauvignon and Merlot as well as White Bordeaux blends and dessert wines. But many wine producing countries from around the world also use this shape. The curve on this bottle near the neck serve as a way to trap sediment in the shoulder of the bottle. When pouring the wine, the sediment is held back by the shoulder, so it won't be transferred into a glass or decanter. This bottle is probably the most common shape for non-sparkling wine bottles on the market.

Burgundy — The Burgundy bottle is a little fuller than the Alsace flutes and has a slightly wider base than the Bordeaux bottles. This bottle shape was created sometime in the 19th century especially for Pinot Noir and Chardonnay wines. But many wine regions in

France and around the world use them for a variety of different grape varietals.

Now that I've put the bottle shapes on your radar, the next thing you might think about is the color of the glass and the punt – or lack thereof – on the bottom of the bottle.

Essentially, tinted wine bottles are used to protect the wine from sun and light exposure. Too much light can start to erode the wine by deteriorating its aromatics, flavors and structures. The colored glass helps protect that from happening. The color is fairly subjective as it's a preference of the winemaker. But sometimes it ties back to a region or as a marketing tool.

Customers tend to like to see the color of their white wine and rosé in the bottle, so many of those are featured in clear glasses. Many red and sparkling wines will feature a medium- to dark-green tinted glass. And some wine will feature a brown-tinted glass from some of the Old-World winemaking regions like France, Germany and Spain.

Then there is always the marketing aspect. Some bottle colors are meant to evoke a certain type of brand recognition or style of wine. When customers see a blue bottle, they associate that with a particular brand or style of wine that is on the sweeter side.

Lastly, getting to the bottom of things, there is the punt discussion. Every now-and-then a student will ask me about the punt or indentation at the bottom of the bottle. It appears that some generalizations exist about the quality or expense of the wine associated with if the bottle has a punt or not.

As with most things in wine, the punt started out of necessity. The glassblowers in the early days of making bottles needed to add an indentation in the bottom for more stability. It appears that if a flat bottom was attempted, any imperfection or miscalculation could create problems with the bottle standing upright. Now it seems like that choice is up to the winemaker – punt or no punt – based on their needs and / or sense of tradition.

: the shopping — alternative packaging

If the thought of purchasing a bottle of wine with a screw cap makes you cringe, the concept of alternative packaging will definitely take some time to get your head around.

In terms of alternative packaging, I'm referring to wine that come in boxes, cans, plastic bottles and kegs. Yes, even kegs.

I mentioned this earlier, but it bears repeating. It is best to keep an open mind, or you might get left behind as things in life change and new technology arises.

The changes in how we entertain, socialize, travel and live our lives in this day and age bring new and interesting ways to engage and interact with wine. Younger consumers, from the youngest of the Millennials to the oldest of the Generation Z generation, have strongly influenced consumer behavior and pop culture.

Love it or hate it! The concept of alternative packaging is not going away any time soon.

I have fond memories of throwing parties in my college years. I would ask people to bring their favorite alcohol and I would supply the wine – a white, red and rosé – in handy boxes of wine for my guests.

It served its purpose back then. It was inexpensive. It was plentiful. It was tasty…enough.

Now it seems we're in an interesting space where the desire for quality wine coupled with the need for winemakers to adjust to our changing lives, have ushered in interesting new options and upgraded takes on older familiar items.

With any business, there will be trends that last and trends that fade. Overtime, we will see how these items fair and keep an eye out for the emerging trends both currently here and on the horizon.

Let's take a quick look at the top two options in the alternative wine package category.

Box Wines / Bag-in-Box (BIB) Wines

Box wines have provided large format wines, typically in the 3 Liters to 5 Liters sizes, for decades now. After years of being associated with "bulk" and "inexpensive" wines, select modern-day winemakers have taken a new liking to the format and use it as an alternative packaging to supplement many of their bottled wine offerings.

Essentially a Box Wine, or also known as Bag-in-Box (BIB) wine, is wine placed in what's called a Polyethylene "bladder" and contained and dispensed from a corrugated, cardboard box. Some boxes are even made of heavy pressed wood.

We have to thank the clever country of Australia for this addition to the wine world. The initial invention was in 1965. But with a few tweaks and a couple years later came the "pièce de résistance:" the plastic, air-tight spout attached to the bladder. This helps keeps oxygen away from the wine, which prolongs the freshness of the wine for up to three weeks.

The goal for many winemakers who now utilize this packaging is to get rid of the stigma that all box wines are generic, bulk wines. These will not be the wines you store in your cellar for years and years. Many of the box wines will actually come with a "use by date." However, the quality of wine found in the boxes has improved over previous years.

The Box Wine concept is regaining favor in countries like Australia, New Zealand and Southern France. It is slowly becoming a growing trend in the United States. Part of its popularity is being pushed forward by Millennials who enjoy more casual approaches to wine. Its lower price points don't hurt either. Others enjoy the fact that Box Wines have a lower Carbon Footprint compared to bottles because they are lighter and easier to transport, helping to reduce carbon-dioxide emissions.

Box wines are ideal for a variety of consumers. It's suited for those who want fresh wine daily, but only drink one glass per night. And then for those on a budget who want to enjoy wine regularly without breaking the bank.

Wine in Cans

The Canned wine segment has been steadily growing in popularity. From the years between 2015 and 2017, it has been reported that sales have more than tripled to what is now a $45 million business.

Like Box Wines, select modern-day winemakers have sought out cans as an alternative package to their bottled wine options. Some companies offer their wine in bottles, boxes and cans. And some companies exclusively offer wine in cans.

Most can wine options come in the 375ML size, which is about 2 1/2 glasses of wine. Other smaller options come in between 250ML and 187ML sizes.

Again, these are not wines for cellaring. These wines are meant to be consumed young and fresh in more casual settings. Some of the traits that attract consumers to can wines is that there is no corkscrew or glassware needed. And the cans are lightweight, easy to transport and travel with, and they dispose of easily when its empty.

: the shopping — online shopping

Living in this ever-changing technological age, the Internet brought a wide range of shopping to our favorite electronic devices some time ago. That also includes shopping on online for wine.

Shopping online for wine is a tremendous opportunity for wine lovers who want to explore wine options found outside of their residential areas. It also proves attractive for those who want the convenience of having wine delivered directly to their home or office.

This accessibility, however, does not come without its caveats.

: Shipping costs

Convenience does often times come at a premium. Wine bottles are both heavy and fragile. A standard-size bottle of wine weights just under 3 pounds. A case of wine, 12 bottles, will weight anywhere between 30-40 pounds. To protect the wine during the shipping process, companies have to invest in special packaging to prevent the wines from any damage that could occur via transport. Lastly, these companies are required to have special licensing in order to legally ship wine around the country. As a result, shipping fees can cost between $35 to $40 per case of wine. Some companies offer shipping specials or memberships to offset or reduce those costs. But getting the convenience of having wine shipped directly to your home, office or local pickup location in your neighborhood might be well worth the expense for many consumers.

: States restrictions on alcohol shipments

While we live in the United States of America, the land of the free, there are still a lot of laws and restrictions that affect our lives on a daily basis. One of those laws, set up by individual State governments, is the restriction of receiving shipments of alcohol from outside of state lines.

A three-tier system of wine and spirits sales was put into place at the repeal of Prohibition in 1933. That legislates how alcoholic beverages can be sold. Be sure to check with retailers, or your State laws, to see if you are able to receive these types of wine shipments. If sending a gift certificate from one of these stores, make sure the state allows the gift recipient to make an order and receive the shipment in their state.

At the time of this book being written, The United States Supreme Court was hearing arguments about the regulation of wine sales, shipping and distribution. A decision is expected to be handed down sometime in 2019.

: the shopping — in-store tastings

Truly learning about wine comes through the practice of tasting wine.

You might have read CHAPTER FOUR and got a clear sense of what Chenin Blanc is as a grape varietal. Without tasting it, however, you don't really know.

Therefore, I strongly suggest that – when at all possible – take a few moments to try the free sample of wine at your farmer's market, wine shop or grocery store.

It's the perfect way to "try before you buy." It is also an excellent opportunity to expand your smell and taste memory, see how a grape varietal differs from region to region or by producer and potentially learn something new from the person offering the sample.

: the serving

You have your wine. You have people coming over. You're excited.

Well, you were excited. Then panic starts to set in.

Will they like the wine? Do I have a wine opener? Do I know how to open the wine while people watch me? Do I have the right glasses? Who do I serve first? How do I handle the situation properly?

First off, it's just wine. Relax. Open a bottle and pour yourself a glass or two. I'm going to go through a few ways, from a serving prospective, to make the situation manageable.

Here are a few tips to help you feel like a sommelier when entertaining.

1. Find a wine key / corkscrew you are comfortable using. Sommeliers tend to love the double-hinged waiters corkscrew.

2. Open the bottle or bottles of wine you will serve. Test the wine to make sure the bottles are free of faults or damage and are suitable for serving.

3. Get a serviette / clean table napkin to carry with you. Wipe the bottle clean of any drops of wine after pouring.

4. Serve clockwise when serving guests at a dinner. Pour for the ladies first along this path. Then circle back to pour for the men.

5. Fill the glasses one-third to half way with wine. This should give you between a standard 4- to 5-ounce pour per person.

6. Position the bottle in your hand so each person can see the label as they are being served.

7. Display bottle nearby or return chilled bottle back to the ice bucket or refrigerator for continued chilling.

See there. Professional at-home wine service in seven simple steps.

: the serving — aerating and decanting

Now you are really excited. Like super-duper excited. This time it's because you get to open up that bottle that has been resting, maturing and waiting for that special occasion.

It's that 1966 Napa Valley Cabernet Sauvignon we've been discussing. When you purchased it, you were instructed by professionals, like me, to decant it and allow it to aerate.

What exactly does that mean? Is that two separate things or does it all happen at the same time?

Many people tend to use the practice of aerating a wine and decanting a wine interchangeably. While these practices do go hand-in-hand at times, they have two different objectives.

Aerating a wine is the practice of allowing it to interact with the oxygen in the air. The aerating process happens when we pour the wine a glass or container. It happens more intently when you swirl the wine in a glass. It continues to happen in your mouth if you try the "slurping" or "trilling" technique.

These aeration practices allow the contents of the bottle to connect with oxygen, helping the wine to oxidize a bit and evaporate a bit.

When that happens, the components of the wine integrate with oxygen and accentuate the varied structures and components of the wine. Not only can you smell the wine's aromatics more effectively, but you can also taste the nuances more closely.

You can aerate select white, red, orange and rosé wines to allow them to stretch and breathe and get comfortable in the glass. Don't aerate your sparkling wine. Drink those immediately while they are bubbly and cold.

Decanting is a bit different. This is the act of removing any sediment from the wine. But while decanting the wine, the aerating process is also happened while you are pouring the wine into a vessel like a carafe or official decanter.

Make sense?

Decanting is more the process of pouring the entire contents of a bottle into a carafe or decanter with the hopes of separating the wine from the sediment. Remember that sediment occurs in red wine, as the wine ages and the tannins soften it forms particles in the bottle called sediment. White wines might have tartrates or wine crystals, but there's no need to filter those out as they might be minuscule in quantity unless it is an older bottle. While doing this, the wine is automatically aerating and interacting with oxygen.

Let's get back to that special bottle we've been teasing for much of this chapter.

We have the 1966 Napa Valley, California Cabernet Sauvignon. It has been resting nicely in the cellar for some time. You'll learn exactly what that entails in the "storing" portion later in this chapter.

It's time to decant this baby.

Pick up the wine very slowly and carefully as not to shake or move the contents of the wine too much. If you happened, by accident to shake it up a bit, just stand the bottle up for about 30 minutes to one hour to let the sediment settle again.

Cut and remove the foil and be sure to wipe the neck clean of any particles from the foil or the cork. This is a great time to test the wine in advance of serving it to others. This is an old wine that has been resting for a long time. It could be passed its prime. It could be corked. It could be damaged by some other wine fault. If the wine is okay, you can start the process of decanting.

Take a light source of choice, a candle, flashlight or phone light, and position the light in front of the bottle where it can clearly light the contents of the bottle with you holding it. The bottle should be in front of you and the light source should be in front of the bottle. Slowly pour the wine into the container of choice. Make sure the container can fit the entire contents of the bottle.

Once you get about half-way through the contents, you will start to see the sediment more. Ideally this bottle of wine has a pronounced shoulder like a Bordeaux-shaped bottle. If so, the sediment will begin to accumulate in the neck of the bottle as you continue to pour very slowly and very carefully.

When you start to see a good amount of sediment, stop pouring. The remaining wine in the bottle will be full of sediment. We want to keep that away from the rest of the wine that is now in the carafe or decanter.

Set the bottle aside to a place where people can review the label. Now, grab your decanting vessel and carefully pour the wine into the glasses.

With a wine of this age, there is no need for much continued aeration. You would want to pour and begin to drink the wine right away. If the wine had some age on it, but wasn't anywhere near 50-plus years old, you can aerate the decanted wine for 30 minutes to an hour in the container for extended contact with oxygen.

If this is all-too-much for you, I have a short cut. It takes out some of the "Pomp and Circumstance" and can be done with a few accessories.

You follow all the steps from opening and dusting off the bottle. Taste it for quality assurance. You can then opt to use a cheese cloth. It has to be something fairly porous that allows for the transfer of liquid. Or you can purchase a wine filter.

Place that "filter" over the mouth of the decanting vessel, secure it firmly in place and pour the wine. The wine will pass through the "filter" and the sediment will be left behind.

: the serving — serving temperature

I think now is a great time for a Pop Quiz.

Come on! We're near the end of the book. It'll just be two questions. I'll even make it multiple choice. You've got this!

Here we go!

Question No. 1: How should white wine be served?
A. Chilled
B. Lukewarm
C. Room temperature

If you answered "A," white wine should be served chilled, you would be correct! Congratulations to those of you who got it right.

Next question!

Question No. 2: How should red wine be served?
A. Chilled
B. Lukewarm
C. Room temperature

If you answered “A,” Red wine should be served chilled, you would be correct! Congratulations to those of you who got it right.

Okay, now that was a bit of a tricky question. I would guesstimate that 50-75 percent of people might get that answer wrong.

Red wine should be served chilled as well!

The people who got the red wine question wrong would probably argue that red wine should be served “at room temperature.”

When serving wine, all wine should be served chilled: red, white, rosé and orange wine. However, they should be chilled in varying degrees according to the style. Sparkling wine, on the other hand, should be served cold.

Way back in CHAPTER TWO, I asked you remember a couple of temperatures. Those numbers coincided with when weather would get warm enough for the vines to start producing fruit. Those number ranges also correspond with the serving temperatures of wine. Do you remember that? Talk about a full-circle moment in the book!

Wine was meant to be served at cellar temperature. This temperature helps to properly show off the wine’s structures and elements. I imagine that somewhere down the road as we progressed as a society, cellar temperature got turned into room temperature in relation to serving wine.

Cellar temperature is a more constant temperature with very little variation. Those temperatures are between 50- and 65-degrees Fahrenheit, with an ideal average of about 55 degrees.

As a result, ideal serving temperatures are:

- Sparkling wines — 40 to 45 degrees (cold, but not freezing)

- Light-bodied white and rosé wines — 50-55 degrees

- Light-bodied red wines — 55-60 degrees
- Medium-bodied wines (red, rosé, white, orange) — 55-60 degrees
- Full-bodied wines (red, white, orange) — 60-65 degrees

I know some of you might still be reeling from learning that red wines should be served chilled. But think of the alternative serving temperate: room temperature.

Room temperature could be anything. In the winter before the heat gets turned on, your home could be around cellar temperature. In the spring time, that room temperature could elevate to about 68 to 75 degrees. The summertime will easily get you anywhere between 80 to 90- plus degrees, especially if you live in a fifth-floor walkup without central air conditioning.

I'm not complaining. I love my apartment. Stairs are good for me!

There's too much variance in the concept of "room temperature." That is particularly true when leaning toward the hotter temperatures. When red wine is served too warm or even – dare I say – teetering on hot, the alcohol, tannins and acidity can come off as very aggressive. We need that cooler temperature to temper the wine for an ideal alignment of all the structures.

Please don't panic! You can absolutely enjoy your red wines at room temperature. We all do. Whether the wine was chilled in advance or not, it will rise in temperature as it sits in the glass, and bottle, over the course of you consuming it.

The key word here is "ideal." It is okay to drink your red wines outside of cellar temperatures. But for maximum enjoyment, cellar temperature is best.

As we well know, it's not common practice for most restaurants or bars to serve red wine chilled. The problem is that there just typically isn't enough space to refrigerate or chill down the red wine inventory. Some places might not even think to chill them in the first place. The white wine, sparkling and rosé bottles tend to take up that space as the cooler temperature is more of a necessity for these wines.

In these settings, you can kindly ask the serving team or bartender to chill your red wine and / or the wine glass before it is served to you.

You can, however, guarantee the ideal serving temperatures at home. All you need is either your refrigerator, a wine refrigerator and / or an ice bucket.

: the refrigerator

This normal, everyday appliance does the trick quite well to chill and store our wines for general, short-term storage. Here is my advice on how to best use your household refrigerator to get your wine to its ideal temperature.

SPARKLING WINE

- Store your sparkling wine in the refrigerator overnight or for a few days

- If possible, lay the sparkling wine on its side to keep the cork moist

- Take the sparkling wine out and serve the sparkling wine immediately

WHITE AND ROSE WINE

- Place your white wine and / or rosé in the refrigerator overnight or for a several hours

- Take the wine out 20 minutes before you plan on serving them and let them rest on a table or counter unopened

- After 20 minutes, open and serve the wine

- To prevent the wine from being too cold, that rest helps it come up in temperature a bit

RED AND ORANGE WINE

- Place your red or orange wine in the refrigerator unopened for 20 minutes

- After 20 minutes, open and serve the wine immediately

- To prevent it from being too warm, that chilling period helps the bottles reach the ideal cellar temperature

That 20-minute window of letting the white and rosé wines warm up and the red and orange wine cool down works for the various body types – light, medium, full – as the wine will chill or warm at different rates and degrees based on its body type.

While the refrigerator is great for helping wines reach their ideal temperatures, it is not great for long-term storage of wine. The humidity is not at the proper percentage for wines with cork stoppers. Not enough humidity can make the cork dry out over time, ultimately spoiling the wine. If the refrigerator is your go-to place for storing your wine, try to store the wine on its side. You want the wine to have direct contact with the cork stopper to keep it moist. But try not to store it there longer than one month. That issue, of course, is not a concern if you have a synthetic cork or screw cap closure.

: the wine refrigerator

The wine refrigerator is one of the best accessories for wine to come along in a very long time. For most households it will be an accessory in the form of a small to medium plug-in unit. For other households, it might come with the home or have been custom built into the home and serves as an official household appliance.

The great thing about wine refrigerators is that most come with temperature controls. The best ones have multiple temperatures. If you have two temperatures to choose from in your unit, store your sparkling, white and rosé at 50 degrees and your red and orange wines at 60 degrees. If one temperature setting is the only option, set it at 55 degrees for all your wine. Some options don't come with an option and therefore the temperature is pre-set by the manufacturer.

You can follow the same guidelines in terms of serving from your wine fridge that were listed above with a traditional refrigerator.

Wine refrigerators are ideal for long-term storage. These devices allow wine to be stored on their sides. And these machines are typically set to maintain the humidity level between 60 to 80 percent, mimicking the environment of a wine cellar.

: the ice buckets

The ice bucket is ideal to chill your sparkling wine to the proper temperature, for parties or larger events, and for a quick chill at the last minute. And, of course, you can use it for your white, rosé, orange and red wines too.

When opting to use an ice bucket, make sure the container has the sufficient depth. You want the water and ice to cover the shoulder of the bottle or the area right below the foil / metal on the neck.

To prepare the ice bath, use an ice and water mixture containing two-thirds of ice to one-third of water. Ice alone will not help to chill the wine. That won't happen until the ice starts to melt, so add water to the ice to expedite the process. Then add a small amount of salt to help dissolve the ice faster. It's the same concept of salting your sidewalks in the winter, the salt breaks down the ice faster.

Sparkling wine bottles are thicker than non-sparkling wine bottles. Therefore, these styles of wines – Cava, Franciacorta, Prosecco, Sekt, Champagne – will take longer to chill. Plan some additional time for these types of wine to help them reach 45 degrees. Warm to lukewarm sparkling wine is not desirable on the palate.

Keep an eye out for your non-sparkling wines on ice – red, white, rosé and orange – to make sure they don't get too cold. We have a tendency to drink our wines too cold sometimes. If they wine is too cold, we don't get full nuance of aromatics and flavor. The cold temperature masks most of the wine's properties.

If the wines do start to get too cold, remove them from the ice bath and let them rest to come up in temperature a bit. Wine doesn't "skunk" like beer, so it is fine to rotate the wine in and out of the ice over a short duration of time to achieve the optimal serving temperature.

: the serving — glassware

Glassware has a charming way of enhancing the wine experience – from the shapes to the stem heights to the brilliant sheen on a freshly polished bulb.

Glassware is definitely important to the serving experience. There are glassmakers that make wine glasses to suit specific types of wines, from sparkling to Pinot Noir to Cabernet Sauvignon to Syrah. These glasses are lovely because the bulbs are crafted and shaped to make the wine touch your palate a certain way in order to best show off the grape varietal.

But drinking a Bordeaux out of a Burgundy glass is not necessarily a deal-breaker. In fact, there are some regions around the world where people drink wine out of tumblers, and they are as happy as can be.

In the wine world, professionals opt to use what we call "A.P." (all-purpose) glasses. An all-purpose wine glass tends to resemble a Bordeaux-style glass. It has a tulip-shaped bulb that can hold 8 ounces of wine or more. We pour just about everything from these glasses including sparkling, red, white, rosé and orange wines.

Ideally you want the glass to be fairly thin and be free of a bump or ridge around the rim of the glass. That sounds very particular, but you just want a clear path for the wine to enter your mouth. And, most importantly, you want to make sure the mouth of the bulb tapers inward. The mouth should be narrower than the center part of the bulb. This allows in the proper amount of air to adequately aerate the wine.

Glasses with stems are ideal for semi-formal to formal occasions. Typically, it is best to hold wineglasses by the stem. That help keeps the wine away from your body heat, which can warm up the wine too much. It also keeps the bulb free of fingerprints as oil and food accumulate.

While stemless glasses are lovely, your fingers repeatedly touch the bulb and can leave tons of marks. These glasses are better suited for more casual situations.

There are lots of great glass options on the market, from the inexpensive to the wildly pricey. The best approach is to select glass options within your budget and personal style parameters.

But please do me a favor! Bypass the plastic cocktail and soda cups for parties and picnics and such. There are lots of great seamless, plastic wine glass options for those times when proper wine glasses – or glass in general – are not viable options.

: the serving — wine with food

Pairing wine with food is one of my favorite things – in both life and in work.

There is this sort of magical thing that happens when the right combination of ingredients is introduced to the right wine. It is the perfect marriage – when paired properly. It's art and alchemy.

In the United States, it is not common, every-day practice to pair wine with food. In fact, sometimes food is an afterthought in our culture. We might go out to Happy Hour and we drink and drink. Then we get to the point where we've had so much wine, that it might be time to get some food.

In the Old World, it is common practice to have wine with food during most situations. The wine enhances the food. The food brings out more in the wine. The two balance each other out. And the overall experience is elevated by the pairing.

Teaching food and wine pairing classes gives me great pleasure because I immediately see the reaction when students understand how the two work together. It makes me feel grateful that I can provide them with a skill to life by incorporating food into their wine consumption patterns on a regular basis. Or vice versa.

There are some guidelines, however, to keep in mind when it comes to properly pairing wine with food.

The number one thing I want you to remember is balance! You don't want the wine to overpower the food and you don't want the food to overpower the wine. With the pairing, you want to create a nice, harmonious balance.

Also, don't lock yourself into the color of the food with the color of the wine you're having – like red wine with red meats and white wine with white meats. It's more about matching textures and balancing flavors.

To help make your food and wine pairing adventures more successful, here are some rules of thumb.

Match Body with Body

An easy way to pair things is to match the body (texture and weight) of the wine with the body of the food. Don't focus on color. You just don't want the items to conflict with each other. You want to create harmony. Therefore, you would not want to serve a nice Pinot Grigio with a barbecue beef brisket. The flavor, fat and protein in the beef would severely overpower the wine. But that white wine would work well with a lighter dish like Soupe de Poisson (fish soup) or Chicken Caesar Salad.

Versatility is Key

You want a wine that is going to be as versatile as possible, whether it's red, white, rosé or orange. That means that the components of the wine – alcohol, acidity, tannins and flavors need to fit with a wide range of dishes.

Watch the Alcohol Content

Higher alcohol wines have a tendency to clash with certain seasonings, spices and flavors. If at all possible, keep your wine choices in the moderate alcohol range, from 11 to 14.5 percent. Save the harder stuff for later after you have a full belly.

Embrace Acidity

High-acid wines are food-loving wines. Acid is a palate-cleanser. It washes your mouth clean of buttery, spicy or meaty flavors, leaving you ready for that next bite! Plus, it gets the appetite going, making you able to finish your plate and maybe even go for seconds.

Watch the Tannins

Tannins are great with fatty foods. But too much tannin can clash with salty items or spicy flavors. Tannins can be soft, moderate or firm. It is the structure in red wines that basically leaves your mouth feeling dry like the moisture has been removed. Therefore, the richer, creamier dishes work best with moderate to firm tannins.

Be careful with Oak

Wines with a heavy oak content can clash with some culinary flavors. Chardonnay can be a tasty, oaky white wine when consumed with meals that have some rich, toasty and creamy components. However, be careful when pairing that with more delicate dishes. For medium-textured dishes, it would be best to select Chardonnay that has light oak or maybe no oak at all. Dishes that are slightly sweet, a bit fruity or tangy can spell trouble when combined with too much oak!

Ultimately the goal is to have fun with both your food and your wine. Experiment with the pairings! Take some risks! Get lost in the process! Sometimes the best way to learn is by learning what does not work.

Bon appétit!

: the storing

Wine can be stored for 20 minutes, 20 days, 20 weeks, 20 months or 20-plus years. It really depends on the wine and your situation.

There are quotes flying around the wine industry that about 60 or more percent of Americans drink their wine within the first 24 hours of purchase.

I'm not sure the accuracy of that number, but I will agree that it is more common for consumers to drink their wine right away as opposed to purchasing for the hopes of cellaring it.

I haven't gotten to the point in my life where I have amassed a collection of wine that I want to cellar and mature. Blame it on the size of my New York apartment or my veracious appetite for wine, but I'm not quite there yet. Most of the wine that comes into my life usually hangs around for about 20 minutes or so before it is opened. That's just me.

But there are a great number of wine lovers who do purchase wines to enjoy down the road.

If you are shopping to collect wines, your eyes are toward the future. You're anticipating an auspicious occasion or want to enjoy something that is well aged. You have carefully researched options. You've made an investment. And you are waiting for that wine to peak to perfection and get a nice return on that investment.

To protect your investment and try to avoid a major disappointment, there are some proactive steps you can take when storing wine. There are tips for the casual collector as well as a more serious enthusiast.

: the storage — short-term

Maybe you don't consider yourself a collector and have no intention of cellaring your wine for several years. But by some way, you have managed to secure some wines that are worth keeping around.

Cellaring wine for the short-term could mean anywhere from a few months to up to five years. You might not want to invest too much money into storing these wines, but you definitely want to protect them.

Here are some dos and don'ts to help you manage the process!

The Dos

- Do find a cool, dark place to store your wines: basement, closet, lower cabinet, storage bin under your bed
- Do store it in a low lying place
- Do store the wine on its side
- Do invest in a small wine refrigerator
- Do look into wine storage companies for multiple bottles you can't manage on your own

The Don'ts

- Don't store on top of the refrigerator
- Don't expose the wine to sunlight
- Don't hang or store wine on high shelves or upper cabinets as heat can damage the wine
- Don't store the wine in the kitchen — this room is a constant source of heat fluctuations
- Don't forget that not all wine is meant to be aged

: the storage — the long-term

This collection of yours will be with you for maybe five, 10, 20-plus years. This is a lot of responsibility. These bottles are your investment and you want to protect them from potentially harmful influences.

You can seek out help to professionally store your wine over time or take on the responsibility yourself.

For long-term storage, you want to create a space that closely resembles the traits of an underground wine cellar where these wines can rest. If you have instructions to drink a wine between 2028 and 2030, you want to do your best to make sure it receives all the care and consideration it needs to mature properly. At the end of the cellaring time, you are hoping that your patience will have paid off and that the wine matured into a complex, dynamic wine.

A little organization goes a long way in managing a wine cellar. It can save you time, energy and money. And it can help prepare you for worst-case scenarios.

A classic wine cellar environment has been proven to be the most effective and ideal way of storing wine for the long haul. The cellar is consistent. The cellar is controlled. The cellar is still.

To mimic those underground cellar conditions in an above-ground location, the following strategies are vital to the process.

SECURE BOTTLES — Plan on the proper shelving or storage bins in which to place your wine. They should be sturdy and prevent the wines from rolling around. Wines should be laid on their sides with the labels facing up. The liquid has to be in direct contact with the cork. This helps to keep the cork from drying out.

SMELLS — While you want to keep your cellar clean, you also want to keep it free of strong orders. That could be from bacteria, pests or even strong cleaning supplies. It is very easy for wine to pick up those aromas if exposed to them on a frequent basis.

INVENTORY — Depending on the size of your storing collection, the inventories should be labeled, numbered and catalogued. An inventory system will help you easily find bottles stored in the cellar. With the labels facing up, you can find and view the wines without causing disruption to other bottles. An inventory list is also perfect for insurance purposes. It comes in handy for those unexpected situations like theft, accidents or massive damage to your cellar.

DON'T BE AFFAID OF THE DARK — Committing to cellaring and storing wine until it matures is not for the faint of heart. And it's definitely not for those who fear the dark. With wine already being a sensitive, living and breathing object, it cannot withstand too much light – particularly sunlight. Be sure to keep your wine away from windows or any constant, powerful sources of light.

CONSTANT TEMPERATURE — With an average temperature of 55 degrees, a cellar is a lot like springtime in the Midwest. It can be a little chilly. Be sure to bundle up with a cozy sweater. The wine, however, will be quite comfortable. This temperature allows the wine to age at a slow and steady pace. If the space is too hot, that starts to accelerate the aging of the wine. Conditions that are too cold could cause the wines to frost or freeze. That could potentially push the cork out some, allowing air to get into the bottle.

HIGH HUMIDITY — This is not like being in the tropics or the Gulf of Mexico. However, a certain amount of humidity – around the 60 to 80 percent mark – is great for the cork. It helps to keep the cork intact, along with resting the bottles horizontally for constant contact with the wine. Humidity levels below 50 percent can lead to the cork drying out, causing spills, oxidation and extreme mold.

BREATHING ROOM — While the cellar has to be dark and cool in temperature, it doesn't mean the room's air has to be stuffy or stale. Proper ventilation is how a wine cellar maintains a consistent temperature. Investing in a cooling unit and exhaust system is necessary for the health of the bottles. It keeps the cool air in and pushes the warm air out.

AVOID DISRUPTIONS — The cellar should be a still, quiet place. It is not the place for a flurry of activity or tomfoolery. There is no need to constantly move or handle the bottles. Too many vibrations and disturbances can really disrupt the wine's rest. Let sleeping bottles lie. And enjoy the fruitful reward of your patience in the years to come.

CHAPTER EIGHT: **the continuing journey**

Life is too short to drink wine you don't like.

It's a truism.

A sad one, but a true one.

I'm not saying that the wine you don't like is *bad* per se. But if you don't like it, that's not a good wine for you.

You can develop a full appreciation for a wine. However, that appreciation doesn't mean it has to be your jam.

It's more about a full-circle approach to understanding the wine world: understanding the land, the grapes, the process. Then you can determine what works for you and what doesn't work.

With this book, I sought to give the best possible "less is more" wine foundation I know how as a wine educator, certified sommelier and writer.

My hope is that I provided a streamlined, understandable and accessible guide to wine appreciation through education, pulling the curtain back in a way that makes it easier and more exciting to incorporate wine into your life.

As this book wraps up, I must leave you with some parting advice as you proceed with your wine adventuring.

1) **Everything must change.** Please don't get stuck in a wine rut. Seasons change. People change. Interests change. Embrace new things as you change and grow as an individual.

2) **Own your choices.** Please be proud of what you like. You, and only you, will truly understand your personal palate. Be secure and confident in what your palate wants. And enjoy it.

3) **Wine doesn't have to be so serious.** Please have fun with wine. It doesn't have to be so traditional. Get a little funky. Get a little weird with the wine choices. Get a little obscure with regions, producers and styles. Go for something that seems crazy. Explore the fringe. Go against the grain. Many great things tend to happen outside of one's comfort zone.

The ride doesn't stop here. This might be the end of the book, but it is just the beginning of a very delicious wine expedition. Creating a close-knit, loving and beneficial relationship with wine is a process.

I hope you will continue the journey. Use this book as a jumping off point toward expanding your wine knowledge. Take these chapters, one-by-one or section-by-section and use them as building blocks to get you where you want to be in terms of your personal wine knowledge.

Wine, just like life, is about constantly learning. Then as you seek to advance your knowledge, check out some amazing resources which cover wine in a lot more detail. Here are some sources that I have come to love over the years.

Pick up "The Wine Bible" by Karen MacNeil. Reading her book in my free time as a newbie wine professional helped push me closer to the desire to learn all I could about wine.

Add "Exploring Wine" by Steven Kolpan, Brian H. Smith and Michael A. Weiss and The Culinary Institute of America to your wine library and future studies. This publication gave me the additional tools to help me garner my sommelier certification.

Then round out your collection with "The Oxford Companion to Wine" by Jancis Robinson. This legendary book is well respected in the wine world and is a strong addition to any growing wine library.

Take wine classes – locally or online – to taste and learn. It is important to learn diverse points of view from other sommeliers and wine educators.

If you want to pursue wine as a career, invest in a certification class or a series of wine classes. If this is an untapped passion of yours, explore it. You might end up making a living doing something you really love.

From here, the path you choose next is up to you. It's like a "Choose Your Own Adventure" book. It's your opportunity to hop around to explore your own interests. Decide what approach works best for your life. Have fun. Taste the wine. Smell the rosé!

In addition to all the continuing education opportunities available, be sure to sample as much as you can. Remember, you can't fully learn about wine without physically experiencing it on your palate.

Alexis Lichine – a wine writer, entrepreneur and a forefather of wine education in the United States in the early 1900s – had some very specific advice about learning about wine.

He simply said: "Buy a corkscrew and use it." He pretty much dropped the "mic" on that one, so there you have it!

Enjoy the process! Enjoy the wine! Enjoy the adventure!

And by all means: Drink life up, as much as possible!

: the gratitude

Words cannot express how grateful I am that you have this book.

If I was in front of you, I would high-five you, kiss you on the cheek or give you a big hug depending on what type of affection you prefer and whatever affection was deemed appropriate for the situation.

I appreciate you. I appreciate your time. And I appreciate you purchasing the book. It warms my heart. You have *no* idea. Your support of this book and of my wine career is a validation of my purpose. It supports me in my personal goal of getting reacquainted with myself as a human, a writer, an educator, a sommelier, a business person and a dreamer.

This isn't goodbye, but "until we meet again." And please know – in your heart of hearts – that as long as I'm breathing, teaching, walking, talking and drinking, I am here for you. As long as I am around, I will be around.

As I mentioned at the beginning of this book, I adore wine. You've probably figured that out for yourselves somewhere along the way.

I honestly would truly love for everyone to adore, understand and appreciate wine as much as I personally do.

: the acknowledgments

The fact that you are reading this book at all – considering that this manuscript has officially gone to press – is by the grace of God. It was also pushed forward by stubborn conviction, a "no-day-but-today" mentality and the love and support of some very important people: some whom I have crossed paths with at various points in my life and many who are still very much a part of it.

There are so many people, places and things that have helped carry me up to this point. Since this is my first book, I'm just going to go big on the "thanks" in this one. I'll try to give thanks in a linear manner, the same way I mapped out this book. I will not use the last names of *most* folks. But you'll know who you are.

This book wouldn't be possible without my Mother, Bunnita, and my Aunt Kemberly who introduced me to my first sip of wine one fateful summer many moons ago. That sparked a life-long love affair with this special juice.

To my early drinking buddies – Lisa, Katrina, Kesha, Chris, Greg, Keary – who helped me to keep my thirst alive for this amazing juice in Milwaukee. While we didn't drink much wine in those days, our early antics unknowingly nudged me more into the wine consumption lane.

My undergraduate college years in Jackson and summers off in Milwaukee were full of exploration and fun. There are far too many names to name in this chapter of my life, but much love goes out to my buddies in Stewart and Dixon Hall, Alexander Hall, The

Student Union, The JSU Class Kings and Queens (I was Mr. Sophomore '94-95), all my Miss JSUs, The Honor's Dorm, The JSU Dance Ensemble, the Mass Communications department, "The Blue & White Flash," W23BC TV News, the house parties at The Advantages, the local television anchors who let me shadow them at work, my community of extended family and friends in New Orleans, family reunion gatherings in Milwaukee and road trips from Atlanta to Houston and Memphis to St. Louis and every place in between.

I will never forget the challenges and growth opportunities that living, working, eating, drinking and glowing up in New Orleans presented. Thank you to my New Orleans people for being uniquely you, "The Times-Picayune," The "Houses" of one or another, The French Quarter, Dillard University, New Orleans East, Metairie, The House of Blues, Snug Harbor and countless house parties from Lake Charles to the West Bank. I would be remised if I didn't mention a few friends by name like Quo Vadis, Philpatrick, Gilbert, Steve, Tameka and the Artist formerly known to me as Nelson.

The graduate school years are really where I reconnected with my love of wine and education. My partners in crime where – and still are – Sherri and Jermaine (aka Jermajesty). As a Jacktown "Three the Hard Way" collective, we had a seemingly all-access pass from the PB and George Street Grocery to Seven and Hamp's Place. And Freelons, duh! The love is real down there and I cannot not mention the loving support of people like Tanya, Roishina, Hester, Shan, Dionne, Dr. Lewis, Stringfellow, The Vanskyvers, New Stage Theatre and all my colleagues and sources at "The Clarion-Ledger." Thanks to everyone who attended my house parties on State Street. And a special thanks to the graduate school at JSU – all my professors and advisors – who helped me leave Jackson, Mississippi proudly with a master's degree.

Post-graduate school life back in Milwaukee brought some unique circumstances and life-changing opportunities. Those were balanced out with a fresh set of new friends, adventures and an opportunity to reconnect with family. Traveling, entertaining clients and wining and dining was all-in-a-day's work in my days in Public Relations. Shouts out to my former colleagues at Laughlin-Constable, Caffeine Communications and Cramer-Krasselt. Socializing was a regular activity in the Third Ward, Brady Street, Downtown, Shorewood and sometimes way out in the "Burbs." Balzac is the best and my all-time favorite wine spot! Countless memories were created with Alida, Belinda, Paul, Chaz, Morris, Kevin, Erin, Cindy, John, Theresa, Esther, John, Kim, Don, Katie, Jenn, Claire, Chuck and the Nelson, Gant and Springfield families. My favorite beverage consumption partner at this time was my beautiful and talented sister Simba. She is the best company, plus she lets me drink most of the wine. Hey Storm! Rest in peace Bruce! I loved my time freelance writing for "INFO" magazine, writing three cover stories for the magazine and keeping my journalism writing skills intact. Thanks to Cuvee Champagne Bar in the Third Ward for my large and very special good-bye party/relocation send off to New York City.

The challenge of all challenges essentially greeted me at the airport gate when I arrived in New York City. The Big Apple of my eye. My favorite place on Earth. It was all a dream, until a nightmarish reality set in quickly as I sought to make a new life here. I encountered so many missed employment opportunities, financial setbacks, stabs in the back and swirling questions of if New York City actually wanted me here. Whatever doesn't kill you makes you stronger like the grape vines. I had to learn that you have to prove to New York City that you belong here in order to gain her respect. Otherwise, she will chew you up and spit you out faster than you can say "If I can make it here, I'll make it anywhere." Around the time things started to balance out a bit, I uncovered a new-found passion for wine and met my new, New York City wine family. This "framily" was truly a gift from Bacchus, the Roman God of wine and revelry. Wine and revelry were had in large doses thanks to Carrie, Cynthia, Kara, Jessica, Gina, Jonathan, Tania, Amanda, Janelle, Julie, KT, Regina, Jessica and a rotating cast of customers and extras on the Upper West Side of Manhattan. This wine journey led me to a serendipitous turn of events that inspired me to become certified as a sommelier through the Sommelier Society of America. This is when I met great new friends like

Abby, Nick, Regino, Danny and tons of wine experts, distributors and my first-ever wine students on the Upper East Side of Manhattan.

Through a long period consisting of twisting and turning events, both high and low, I landed and jumped around from places like Astor Wines in Greenwich Village, BTL in Harlem, Adega Wine & Spirits in Astoria, Queens, and NYVintners in TriBeCa. Thanks to the following people, I was able to come out of these experiences and life's challenges a lot stronger than when I entered: Gabby, Hughes, Rodrigues, Colin, Ali, Or, Silvi, Diony, Anthony, Leah, Hillary, Trinity, Wueder, Simone, Shane, Ryan S., Oz, Narisa, Susan, Kara, Ed, Ryan T., Jessica, Suzanne, Jon, John, Derrick, Jose, Kelvin, Arturo, Fernando, Juan Carlos and the crew of NYVintners porters. Then there are my other teams at Taste Wine Co. and Corkbuzz, like Gary, Kristina, Laura, Isaiah, Abe, Shelby, Ehron, Andrew, Amber, Ryan, Dorian, Jake and Jamie. Through these years, I've learned so much more about wine, teaching food and wine pairings, service, friendship, strength in diversity, patience, hustle and that common thread that ties us all together in our humanity. These varied, and sometimes extreme, experiences really shaped me into becoming the sommelier that I am today.

This book was inspired, literally, by the thousands of students I taught over the years at Serendipity, BTL Harlem, NYVintners, Adega Wine & Spirits, Corkbuzz Wine Studio, Taste Wine Co, Pop-Up locations, at-home classes and Harlem Wine Gallery. There are way too many students to name, but all are equally important to me. You have helped shape me into the wine educator I am today. Shout out to my teams at Wine.com, Louis Vuitton Moet Hennessy (LVMH)/Strategic Group!

Inspiration, for me, comes in many forms, shapes, mediums and packages. The "SOMM" movies – from the first documentary to the most recent third installment – have been part of my professional wine life from the time I got certified in 2012 until this book was being finalized in 2018. Thank you for offering up tons of knowledge, shared struggles and introducing me to an incredible cast of professionals. One of which, I consider my "unofficial" mentor and friend, Mr. DLynn Proctor.

Thank you to these very special and talented artists I have in my life that bring me constant inspiration with their songs, performances and spirits: Alyson Williams, Anika Noni Rose and Syndee Winters. I'm so thankful for jazz, R&B/soul, Broadway, documentaries and PBS for providing countless hours of inspiration and positive energy.

Speaking of energy, I can't say enough how the energy of New York City has fueled my passion for chasing all these different endeavors I dream up. This is especially true of the great people of Harlem that inspire me and encourage me on the regular. Thank you to Janelle, Vonnie and Maxwell! Miss Sohad! I love you! I miss you not living up the avenue from me anymore. Hi Auntie Victoria Horsford! And Rest in Peace to my cousin and former Harlemite Giselle King-Porter. Also, thank you to my many, many friends and family scattered all around the world!

A special thank you to everyone who has worked on the book from photography to design, editing to production – Curtis, Anne and a series of freelance professionals. I think it looks beautiful.

This next one is very special for me! I would not have been able to put these thoughts that have been swirling around my head down into these 250+ pages without my literary God Father: Mr. Langston Hughes. Your images are sprinkled around my room. The smell of the Langston inspired candle fills up my work/writing space. I feel your presence nearby where your ashes are buried at the Schomburg Center for Research in Black Culture. And the ability to step foot into your former residence, The Langston Hughes House in Harlem, and feel your energy is beyond comprehension. You were the reason I became a writer, wanted to attend a Historically Black College and University (HBCU) and the reason I fell in love with Harlem way before I've ever set foot in the neighborhood.

While this book was a labor of love, this would not be a reality right now without the generous donations that come through to help me fund this project. My Angel investors. Your support of me and faith in me – without seeing a single word on a page from most supporters – means the absolute world to me.

I can't thank you enough, therefore, I wanted to put in a special thank you here by order of donations. Thank you Bunnita Gant (Mom), Simba Gant (Sister), Storm Gant (Brother), Kara Joseph, Keith Gardner, Maxine Lee, Hester Collins, Nicole Benoit, Alissa Reiher, Carrie Dykes, Abbey Creek Vineyard (Bertony Faustin), Marisa Jane Lyman, Paul Grant, Shannon Westfall, Richard Golaszewski, Suzanne Tran, Narisa Gaffoor, Vijay Gaffoor, Ryan Smith, Joy Lindsey, Jason Oliva, Holly Roloff, Riva Brown, Tammy Yates, Cecilia Gilbert, Marion Gottschalk, Katja Dehaney, Shelley Thomas, Cassie Oden, Jennifer Hergert, Annie Lettenberger, Jamie Metzgar, Joe Kennedy, Susan Ellis, Charles "Chuck" Sanchez, Anne Woelfel, Lisa G, Kemberly Springfield, Christopher Springfield, Anika Rose, Jose Rodriguez, Alex Keahan, Tasha Lloyd, Anthony Coppola, Edwin Aristor, Pamela Berry, KT Goldthorpe, Sherri Williams, Jermaine Alexander, Julie Garman, John Hale, Valarie McCubbins, Alesha Russey and Jeff and Kelly Springfield.

These donations have not just helped me keep the process moving in terms of expenses associated with publishing a book through my company, it also helped lift my spirits and lift my hands to the keyboard at times when I was tired or discouraged or questioning if this book would matter to anyone in the end. Your taking time out to donate meant that this book matters to other people besides just me. The thought of that is what continued to propel me forward through this long process when self-motivation wasn't enough. I knew if I can at least get this finished product in the hands of the people who helped me, then it would be worth all the hard work and sacrifice.

At the end of the day, this book belongs to all of us. You have helped shape and support who I am as a man, writer, sommelier, wine educator, entrepreneur, dreamer, New Yorker and optimist.

I'm almost done.

I would not be able to close out these acknowledgments without saluting this growing voice of diversity in wine, the "leaders of the new school of wine education," my fellow black wine professionals. The sommeliers. The winemakers. The taste makers. The grape growers. The chefs. The Podcast show hosts. The Influencers. There's a movement bubbling up! Sab, you really started something. It's our time. Let's seize the day, make some noise and blow some minds. I love you all and I'm very proud of the work each and every one of you are doing around the country and around the world. Cheers to us!

And last, but certainly, not least, I want to thank *me*. Snoop Dogg, you really inspired me with your Hollywood Walk of Fame speech. So here I go!

I want to thank me for allowing myself to believe that this book was possible. I want to thank myself for making tons of lifestyle changes and sacrifices to plan this out, construct it, write it, revise it, scrap it and start all over again until I got it as close to right as I possibly could. I want to thank me for not letting distractions get the better of me. I want to thank me for not letting the energy of toxic people bring me down. I want to thank me for protecting my creative space and writing time. I want to thank me for making me proud of me again. I want to thank me for taking an active role in shaping my destiny – the way I envision it playing out. I want to thank me for not giving up even when it seemed like so many universal forces were trying to force me to quit. I want to thank me for constantly inspiring me to be a better man. I want to thank me for the motivation to continue to create, grow, inspire, build, accomplish and learn. I thank me for being me. I love you, meaning me, very much!

Now back to thanking everyone else again. I hope this book makes you as proud as it makes me. I'm excited to share this book with the world and continue working on being a published author, inspiring others and encouraging people to get out of their own way and "drink life up, as much as possible."

I thank you! I love you! I hope to see as many of you as I can, as soon as I can, and as often as I can.

~ Charles Springfield

: the author

Charles Dion Springfield is a Certified Sommelier in New York City where he developed a tailor-made lifestyle marketing company called "The Life Stylings of Charles Springfield." The company focuses on his approach to wine, events and joie de vivre (joy of life). The three major divisions include: The Wine Stylings, The Editorial Stylings and The Event Stylings.

He specializes in matching the right wines with the right people and occasions, teaching wine classes, hosting wine events and promoting overall wine appreciation through education.

Prior to falling head first into the wine business, he was an award-winning journalist and public relations executive for nearly two decades. As a print journalist, Charles began his career as a general assignment reporter at the Pulitzer-prize winning "Times-Picayune" daily newspaper in New Orleans. He has written columns and covered everything from crime and health to entertainment and lifestyle feature stories at "The Clarion-Ledger."

In the public relations field, he worked primarily in the consumer marketing / lifestyle realm at three advertising agencies and one boutique public relations firm.

For nearly ten years now he has used his unique approach to marketing communications, his "life styling" experience and love of wine to provide guidance on appreciating, understanding and enjoying wine in ways people never imaged possible.

Charles holds a Bachelor and a Master of Science degree in Mass Communication, both obtained from Jackson State University in Jackson, Mississippi. He is also certified by the Sommelier Society of America which was founded in 1954 and based in New York City.